The Complete Guide to

BUILDING CLASSIC BARNS, FENCES, STORAGE SHEDS, ANIMAL PENS, OUTBUILDINGS, GREENHOUSES, FARM EQUIPMENT, & TOOLS

A Step-by-Step Guide to Building Everything You Might Need on a Small Farm — With Companion CD-ROM

THE COMPLETE GUIDE TO BUILDING CLASSIC BARNS, FENCES, STOR-
AGE SHEDS, ANIMAL PENS, OUTBUILDINGS, GREENHOUSES, FARM
EQUIPMENT, & TOOLS: A STEP-BY-STEP GUIDE TO BUILDING EVERY-
THING YOU MIGHT NEED ON A SMALL FARM — WITH COMPANION CD-
ROM

Copyright © 2012 Atlantic Publishing Group, Inc.
1405 SW 6th Avenue • Ocala, Florida 34471 • Phone 800-814-1132 • Fax 352-622-1875
Website: www.atlantic-pub.com • E-mail: sales@atlantic-pub.com
SAN Number: 268-1250

Library of Congress Cataloging-in-Publication Data

The complete guide to building classic barns, fences, storage sheds, animal pens, outbuildings,
greenhouses, farm equipment, & tools : a step-by-step guide to building everything you might need on a
small farm : with companion CD-ROM.
     p. cm.
  ISBN 978-1-60138-372-3 (alk. paper) -- ISBN 1-60138-372-X (alk. paper) 1. Farm buildings--Design and
construction. 2. Farm buildings--Designs and plans.
  TH4911.C57 2012
  690'.537--dc23
                         2012000679

Printed in the United States

PROOFING: Gretchen Pressley • gpressley@atlantic-pub.com
INTERIOR DESIGN: Jackie Miller • millerjackiej@gmail.com
FRONT COVER DESIGN: Meg Buchner • megadesn@mchsi.com
BACK COVER DESIGN: Jackie Miller • millerjackiej@gmail.com

Printed on Recycled Paper

A few years back we lost our beloved pet dog Bear, who was not only our best and dearest friend but also the "Vice President of Sunshine" here at Atlantic Publishing. He did not receive a salary but worked tirelessly 24 hours a day to please his parents.

Bear was a rescue dog who turned around and showered myself, my wife, Sherri, his grandparents Jean, Bob, and Nancy, and every person and animal he met (well, maybe not rabbits) with friendship and love. He made a lot of people smile every day.

We wanted you to know a portion of the profits of this book will be donated in Bear's memory to local animal shelters, parks, conservation organizations, and other individuals and nonprofit organizations in need of assistance.

– *Douglas & Sherri Brown*

PS: We have since adopted two more rescue dogs: first Scout, and the following year, Ginger. They were both mixed golden retrievers who needed a home.

Want to help animals and the world? Here are a dozen easy suggestions you and your family can implement today:

- *Adopt and rescue a pet from a local shelter.*
- *Support local and no-kill animal shelters.*
- *Plant a tree to honor someone you love.*
- *Be a developer — put up some birdhouses.*
- *Buy live, potted Christmas trees and replant them.*
- *Make sure you spend time with your animals each day.*
- *Save natural resources by recycling and buying recycled products.*
- *Drink tap water, or filter your own water at home.*
- *Whenever possible, limit your use of or do not use pesticides.*
- *If you eat seafood, make sustainable choices.*
- *Support your local farmers market.*
- *Get outside. Visit a park, volunteer, walk your dog, or ride your bike.*

Five years ago, Atlantic Publishing signed the Green Press Initiative. These guidelines promote environmentally friendly practices, such as using recycled stock and vegetable-based inks, avoiding waste, choosing energy-efficient resources, and promoting a no-pulping policy. We now use 100-percent recycled stock on all our books. The results: in one year, switching to post-consumer recycled stock saved 24 mature trees, 5,000 gallons of water, the equivalent of the total energy used for one home in a year, and the equivalent of the greenhouse gases from one car driven for a year.

TABLE OF CONTENTS

CHAPTER 1: DAIRY FARM BUILDINGS & EQUIPMENT 9

THE COMPLETE GUIDE TO BUILDING CLASSIC BARNS, FENCES, STORAGE SHEDS,
ANIMAL PENS, OUTBUILDINGS, GREENHOUSES, FARM EQUIPMENT, & TOOLS

chapter 1

DAIRY FARM BUILDINGS & EQUIPMENT

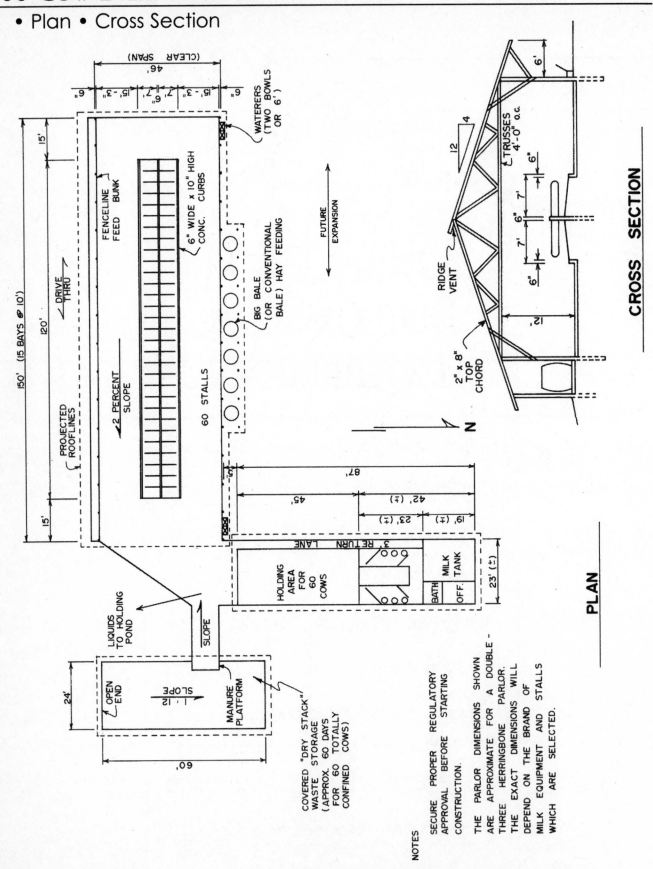

• Plan

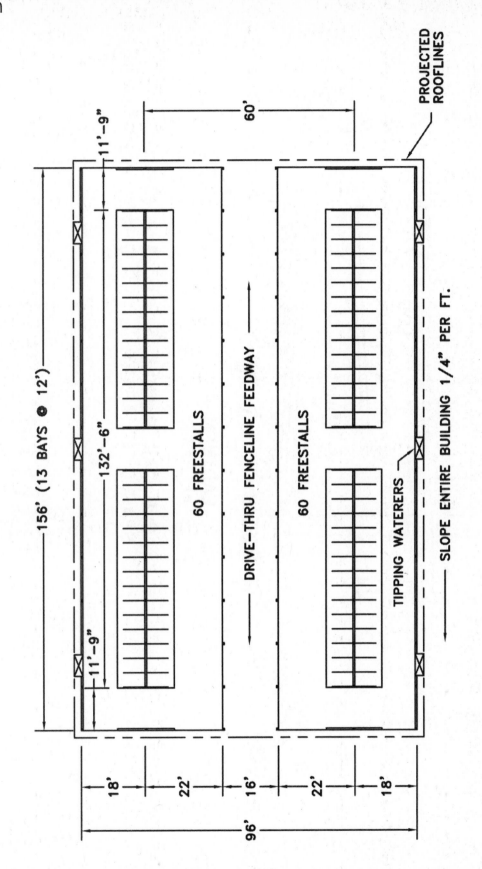

• Freestall Detail

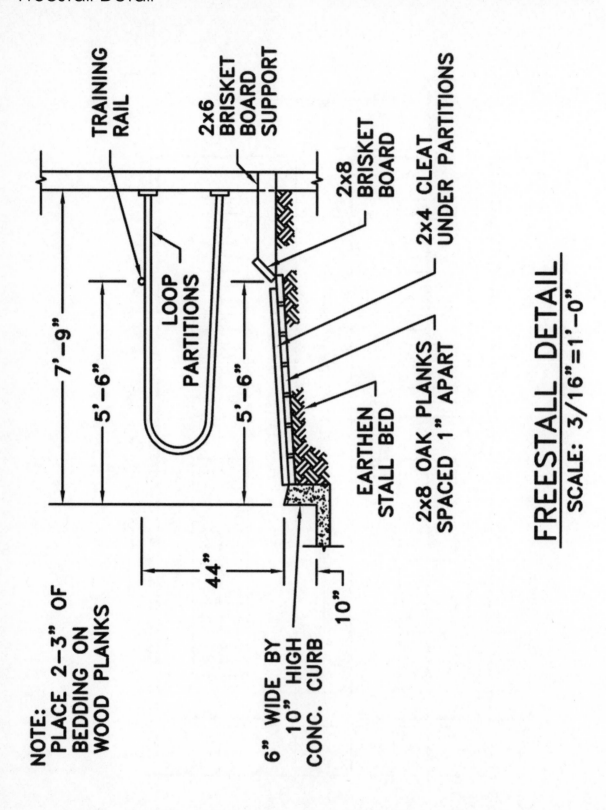

TRAINING RAIL

2x6 BRISKET BOARD SUPPORT

2x8 BRISKET BOARD

2x4 CLEAT UNDER PARTITIONS

LOOP PARTITIONS

7'-9"

5'-6"

5'-6"

EARTHEN STALL BED

2x8 OAK PLANKS SPACED 1" APART

44"

10"

NOTE: PLACE 2-3" OF BEDDING ON WOOD PLANKS

6" WIDE BY 10" HIGH CONC. CURB

FREESTALL DETAIL

SCALE: 3/16"=1'-0"

• Cross Section

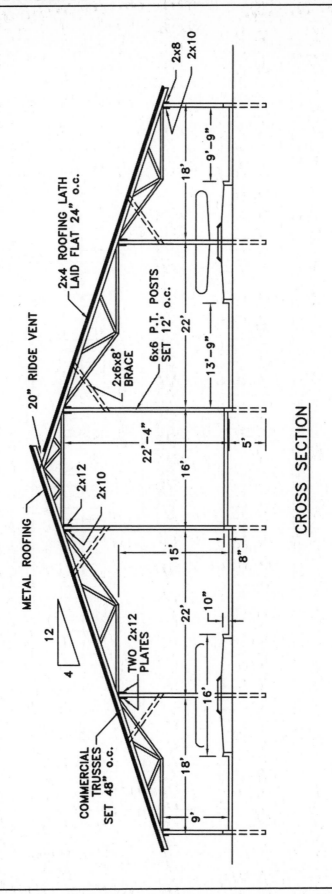

CROSS SECTION

• Plan

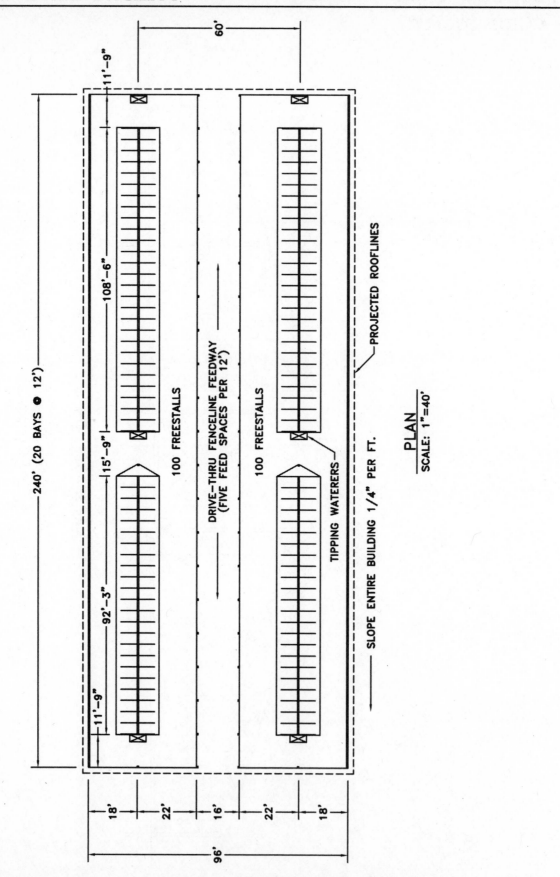

PLAN
SCALE: 1"=40'

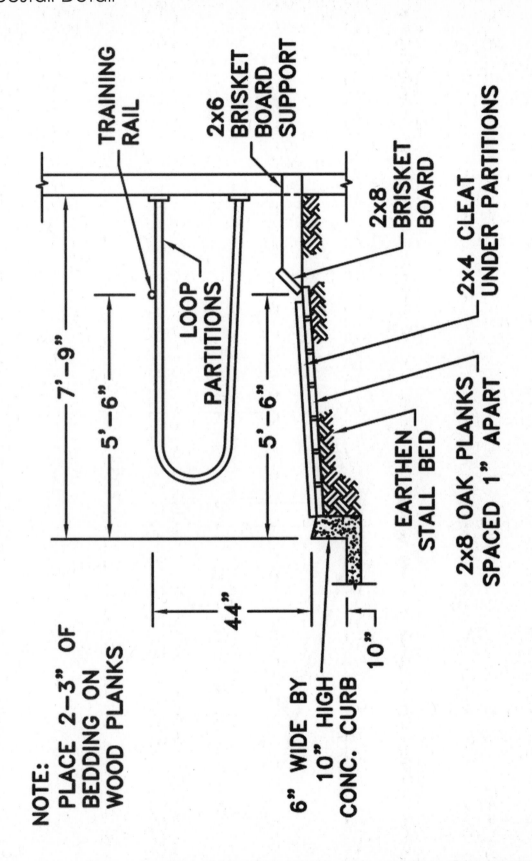

TRAINING RAIL

2x6 BRISKET BOARD SUPPORT

2x8 BRISKET BOARD

2x4 CLEAT UNDER PARTITIONS

LOOP PARTITIONS

EARTHEN STALL BED

2x8 OAK PLANKS SPACED 1" APART

7'–9"

5'–6"

5'–6"

44"

10"

6" WIDE BY 10" HIGH CONC. CURB

NOTE: PLACE 2–3" OF BEDDING ON WOOD PLANKS

• Cross Section

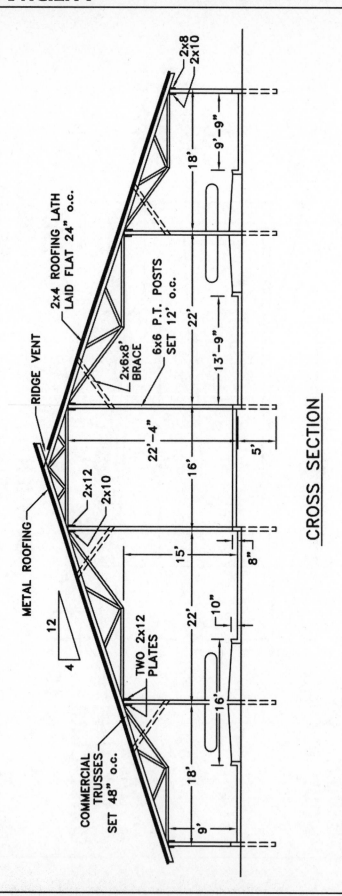

CROSS SECTION

• Floor Plan • Herd Sizes and Building Dimensions

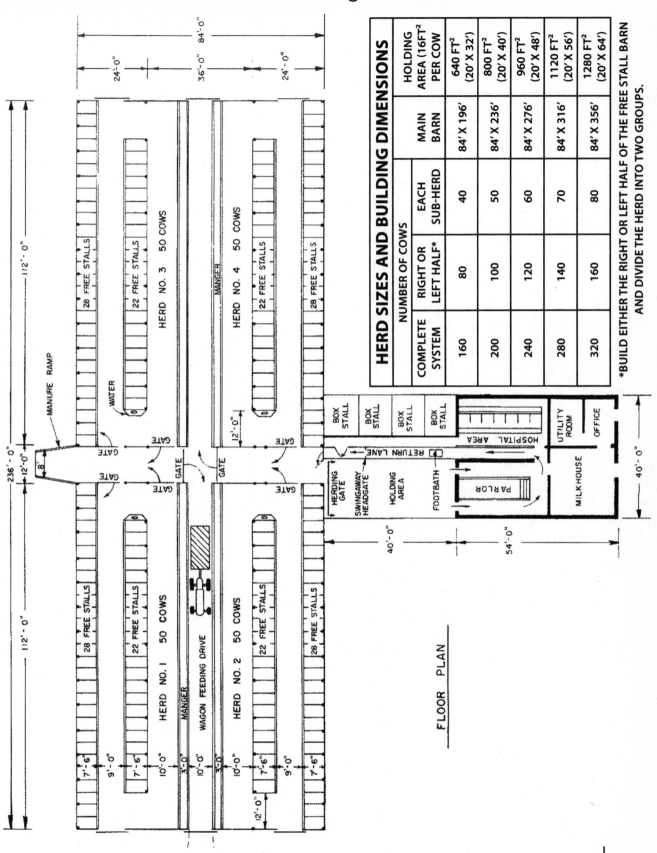

HERD SIZES AND BUILDING DIMENSIONS

NUMBER OF COWS			MAIN BARN	HOLDING AREA (16 FT²) PER COW
COMPLETE SYSTEM	RIGHT OR LEFT HALF*	EACH SUB-HERD		
160	80	40	84' X 196'	640 FT² (20' X 32')
200	100	50	84' X 236'	800 FT² (20' X 40')
240	120	60	84' X 276'	960 FT² (20' X 48')
280	140	70	84' X 316'	1120 FT² (20' X 56')
320	160	80	84' X 356'	1280 FT² (20' X 64')

*BUILD EITHER THE RIGHT OR LEFT HALF OF THE FREE STALL BARN AND DIVIDE THE HERD INTO TWO GROUPS.

FLOOR PLAN

• Cross Section

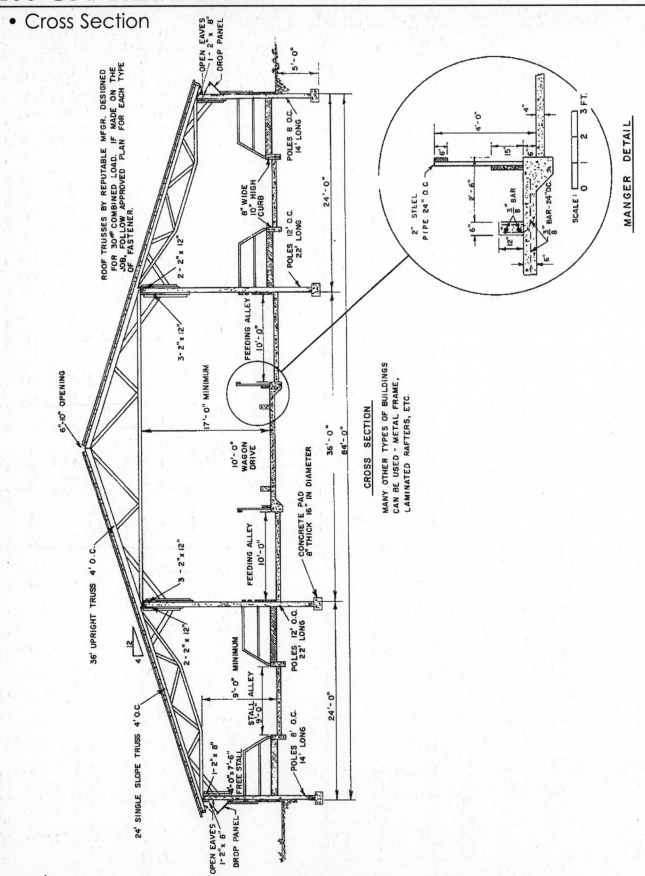

ROOF TRUSSES BY REPUTABLE MFGR. DESIGNED FOR 30# COMBINED LOAD. IF MADE ON THE JOB, FOLLOW APPROVED PLAN FOR EACH TYPE OF FASTENER.

OPEN EAVES 1-2"x 8"
DROP PANEL
5'-0"
POLES 8' O.C. 14' LONG
8" WIDE 10" HIGH CURB
POLES 12' O.C. 22' LONG
24'-0"
2-2"x 12"
FEEDING ALLEY 10'-0"
3-2"x 12"
17'-0" MINIMUM
6"-10" OPENING
10'-0" WAGON DRIVE
CONCRETE PAD 8" THICK 16" IN DIAMETER
36'-0"
84'-0"

CROSS SECTION

MANY OTHER TYPES OF BUILDINGS CAN BE USED - METAL FRAME, LAMINATED RAFTERS, ETC.

36' UPRIGHT TRUSS 4' O.C.
12
4
24' SINGLE SLOPE TRUSS 4' O.C.
3 - 2"x 12"
2 - 2"x 12"
9'-0" MINIMUM
FEEDING ALLEY 10'-0"
STALL ALLEY 9'-0"
POLES 12' O.C. 22' LONG
24'-0"
OPEN EAVES 1-2"x 8"
DROP PANEL
1-2"x 8"
4'-0"x 7'-6" FREE STALL
POLES 8' O.C. 14' LONG

2" STEEL PIPE 24" O.C.
4'-0"
4"
2'-6"
15"
6"
6"
6"
3" BAR 8
3" BAR 8
3" BAR-24" O.C.
12"
6"

SCALE: 0 1 2 3 FT.

MANGER DETAIL

- Interior Pole Bracing and Rafter Detail

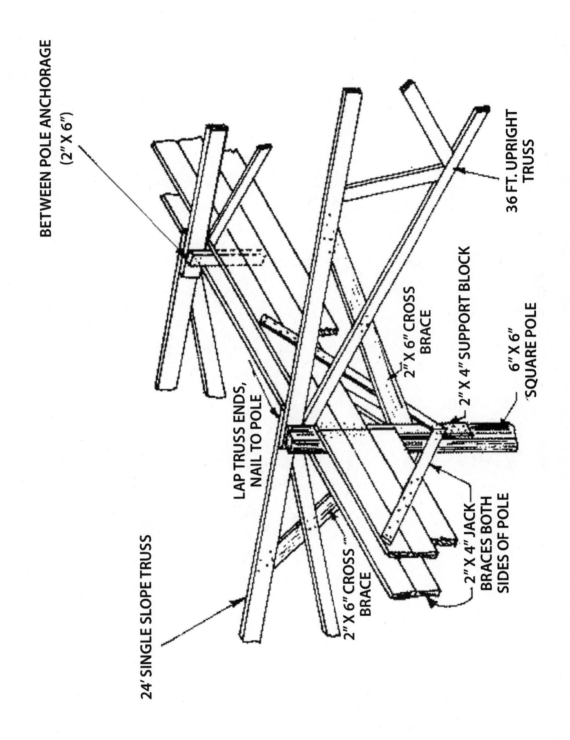

BETWEEN POLE ANCHORAGE
(2" X 6")

36 FT. UPRIGHT TRUSS

2" X 6" CROSS BRACE

2" X 4" SUPPORT BLOCK

6" X 6" SQUARE POLE

LAP TRUSS ENDS, NAIL TO POLE

2" X 4" JACK BRACES BOTH SIDES OF POLE

2" X 6" CROSS BRACE

24' SINGLE SLOPE TRUSS

• Roofing Detail with Purlins and Metal Roofing

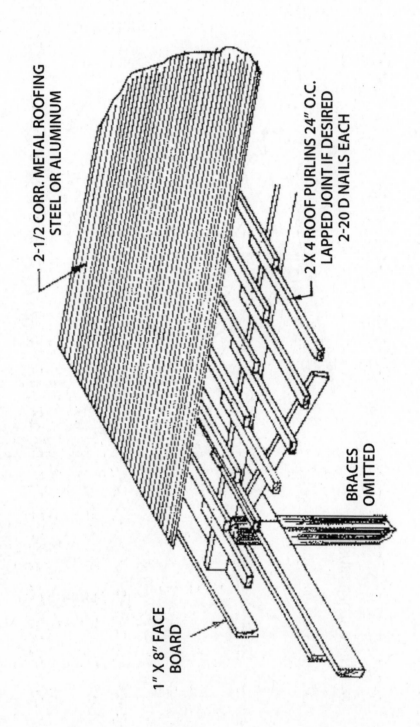

2-1/2 CORR. METAL ROOFING STEEL OR ALUMINUM

2 X 4 ROOF PURLINS 24" O.C. LAPPED JOINT IF DESIRED 2-20 D NAILS EACH

BRACES OMITTED

1" X 8" FACE BOARD

• Open Ridge Vent Detail

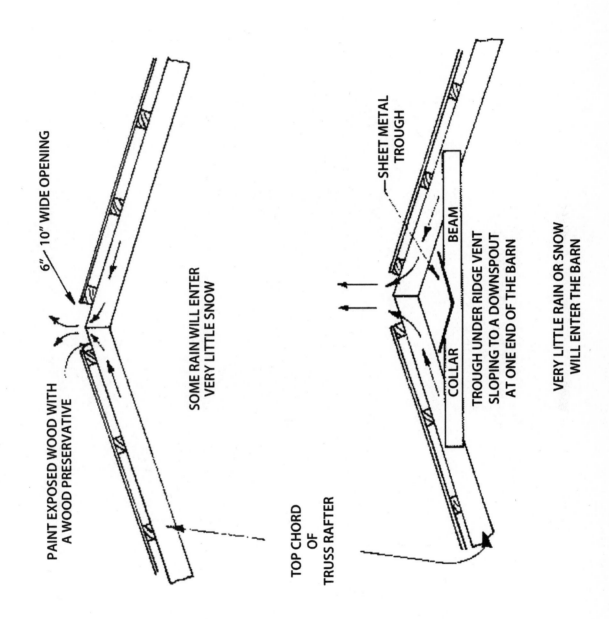

6" 10" WIDE OPENING

PAINT EXPOSED WOOD WITH A WOOD PRESERVATIVE

SOME RAIN WILL ENTER VERY LITTLE SNOW

TOP CHORD OF TRUSS RAFTER

SHEET METAL TROUGH

BEAM

COLLAR

TROUGH UNDER RIDGE VENT SLOPING TO A DOWNSPOUT AT ONE END OF THE BARN

VERY LITTLE RAIN OR SNOW WILL ENTER THE BARN

• Side Wall Detail

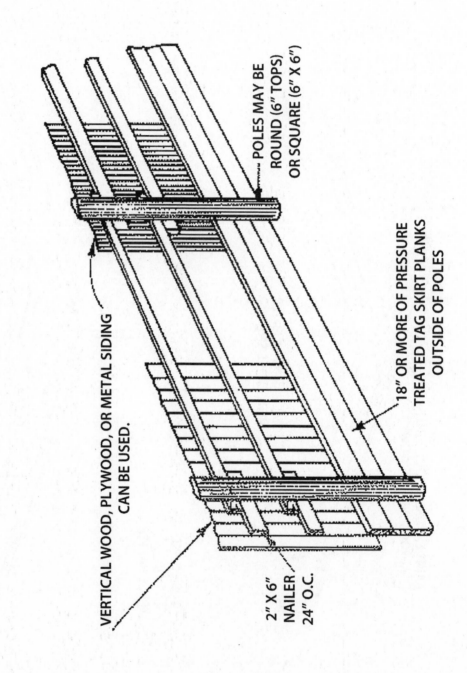

POLES MUST BE PRESSURE TREATED FOR 50-YEAR LIFE. UNTREATED OR COLD-TREATED POLES WILL HAVE A SHORT LIFE. POLES MAY BE ROUND (6" TOPS) OR SQUARE (6" X 6").

POLES MAY BE ROUND (6" TOPS) OR SQUARE (6" X 6")

VERTICAL WOOD, PLYWOOD, OR METAL SIDING CAN BE USED.

18" OR MORE OF PRESSURE TREATED TAG SKIRT PLANKS OUTSIDE OF POLES

2" X 6" NAILER 24" O.C.

• Side Wall Pole Bracing and Nailing Detail

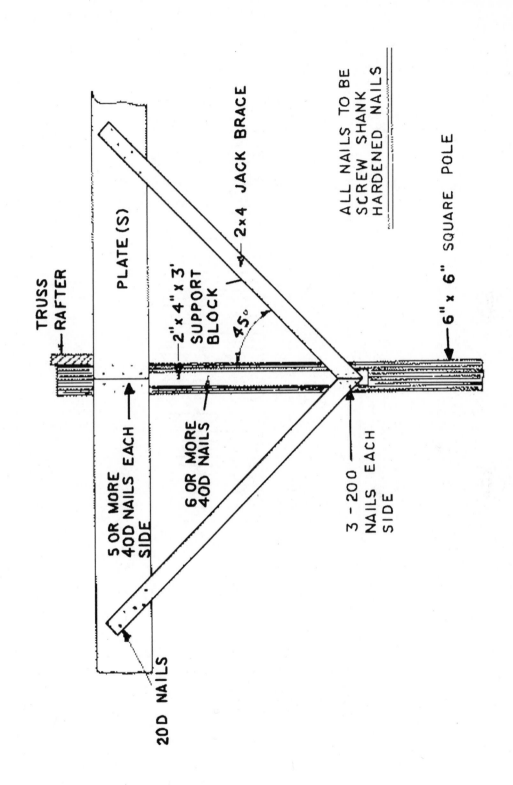

TRUSS RAFTER

PLATE (S)

2"x4" JACK BRACE

2"x4"x3' SUPPORT BLOCK

45°

ALL NAILS TO BE SCREW SHANK HARDENED NAILS

6"x 6" SQUARE POLE

5 OR MORE 40D NAILS EACH SIDE

6 OR MORE 40D NAILS

3 - 200 NAILS EACH SIDE

20D NAILS

• Panel Window Detail

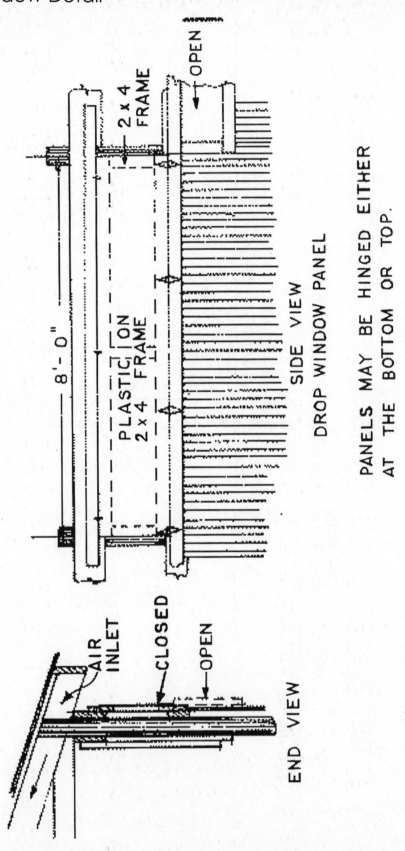

200 Cow Freestall Barn

- Alternate Location for the Milking Center, Holding Area, and Hospital Area

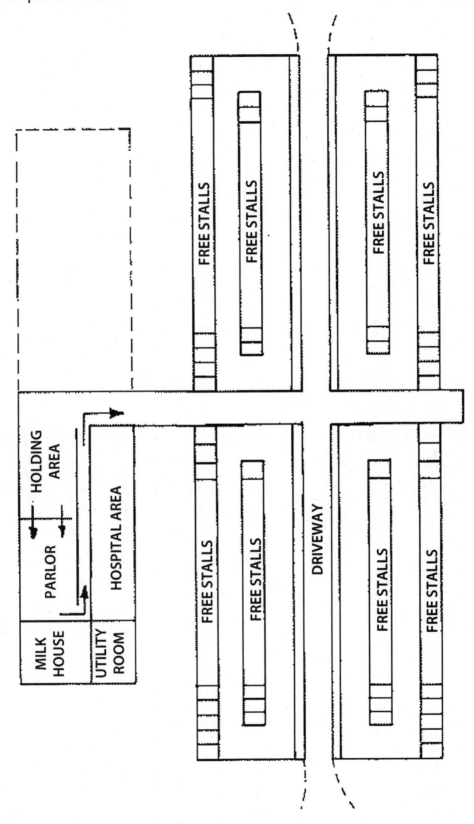

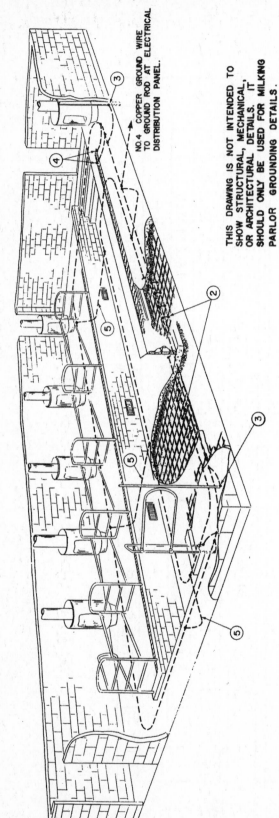

NO.4 COPPER GROUND WIRE TO GROUND ROD AT ELECTRICAL DISTRIBUTION PANEL.

THIS DRAWING IS NOT INTENDED TO SHOW STRUCTURAL, MECHANICAL, OR ARCHITECTURAL DETAILS. IT SHOULD ONLY BE USED FOR MILKING PARLOR GROUNDING DETAILS.

NOTES:

1. NOTE ABOUT EFFECTS OF AN EQUIPOTENTIAL PLANE: IF THE FLOOR IS AT THE SAME POTENTIAL AS THE CONDUCTIVE EQUIPMENT AND ANY STRUCTURES ACCESSABLE TO THE ANIMAL, STRAY VOLTAGE PROBLEMS CANNOT EXIST. THIS CAN BE ACCOMPLISHED BY PLACING A BONDED NETWORK OF WELDED WIRE MESH IN THE FLOOR. ANY ANIMAL STANDING ON A FLOOR CONTAINING A PROPERLY INSTALLED EQUIPOTENTIAL PLANE WILL HAVE ALL POSSIBLE CONTACT POINTS AT OR VERY NEAR THE SAME POTENTIAL.

2. BOND NO.4 COPPER WIRE TO 6" x 6" (UP TO 12" x 12" IS OK) WIRE MESH IN CONCRETE FLOOR, 2 OR 3 TIMES PER EACH CONT. SECTION OF MESH.

3. IF THERE IS A POSSIBILITY FOR ENERGIZING DUE TO A LOCAL ELECTRICAL FAULT SUCH AS ELECTRIC EQUIPMENT NOT BEING PROPERLY GROUNDED, THEN STEEL POSTS, GATE POSTS, SUPPORT POSTS, FEEDER BRACKETS, ETC. SHOULD BE GROUNDED ACCORDING TO THE NATIONAL ELECTRIC CODE.

4. ANGLE IRON GRATE SUPPORTS FOR FLOOR DRAINS TO BE BONDED BY RESISTANCE WELDING AT 2 OR MORE POINTS FOR EACH CONT. SECTION.

5. 6" x 6" WIRE MESH ON FLOOR OF PIT CONNECTED AT 2 POINTS TO MESH IN COW PLATFORM FLOOR.

6. WIRE MESH TO HAVE A PROTECTIVE CONCRETE COVER TO PROTECT MESH FROM CORROSION OR DAMAGE (1" IS SUFFICIENT, BUT 1 1/2" IS BETTER).

7. 1/4" ROUND STEEL ROD WELDED (RESISTANCE OR EXOTHERMIC) TO FEEDER, EXTENDING DOWN TO WIRE MESH. WELD OR BRAZE 1/4" ROD, MESH, AND NO.4 COPPER GROUND WIRE TOGETHER. INSTALL 2 RODS PER SIDE OF PARLOUR PROVIDED THAT ALL FEEDERS ARE INTERCONNECTED BY METAL PARTS. NOTE: CONDUCTORS AND CONNECTIONS SHOULD BE LOCATED WHERE THEY CANNOT BE EASILY DAMAGED OR DISTURBED.

8. UNDER SERIOUS CONDITIONS, AN ANIMAL MAY BE EFFECTED WHEN IT STEPS ONTO THE EQUIPOTENTIAL PLANE FROM AN AREA BEYOND THE PERIMETER. TO PROVIDE A MORE GRADUAL CHANGE IN VOLTAGE, A VOLTAGE RAMP SOULD BE INSTALLED AT THE ENTRANCE OR EXIT OF THE EQUIPOTENTIAL PLANE. (SEE VOLTAGE RAMP DETAIL)

MILKING PARLOR GROUNDING DETAILS

- Cross Section • Optional Technique for Bonding Parts
- Voltage Ramp Detail

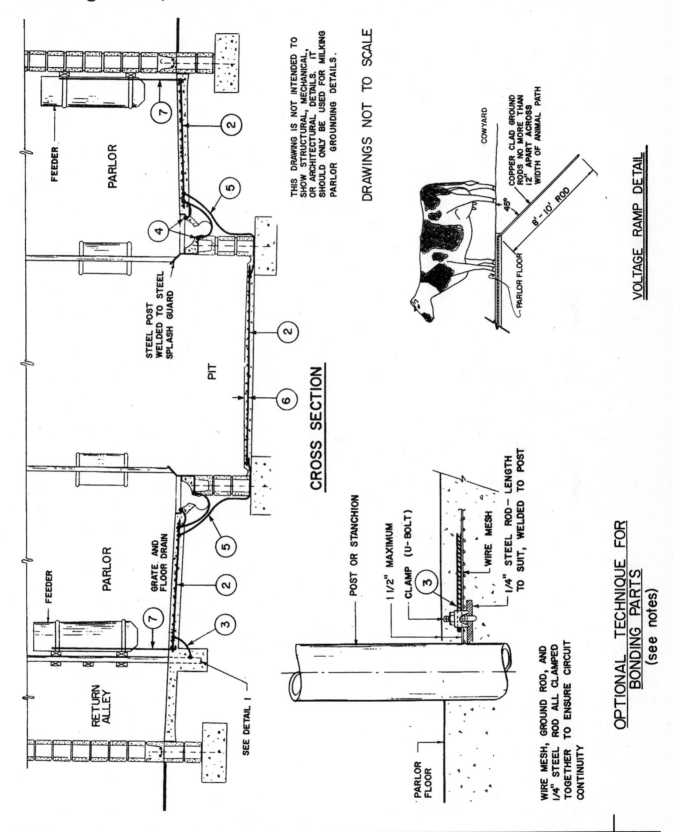

THIS DRAWING IS NOT INTENDED TO SHOW STRUCTURAL, MECHANICAL, OR ARCHITECTURAL DETAILS. IT SHOULD ONLY BE USED FOR MILKING PARLOR GROUNDING DETAILS.

DRAWINGS NOT TO SCALE

FEEDER

PARLOR

STEEL POST WELDED TO STEEL SPLASH GUARD

PIT

CROSS SECTION

FEEDER

PARLOR

GRATE AND FLOOR DRAIN

RETURN ALLEY

SEE DETAIL I

COWYARD

COPPER CLAD GROUND RODS NO MORE THAN 12" APART ACROSS WIDTH OF ANIMAL PATH

45°

8' - 10' ROD

PARLOR FLOOR

VOLTAGE RAMP DETAIL

POST OR STANCHION

1 1/2" MAXIMUM

CLAMP (U-BOLT)

WIRE MESH

1/4" STEEL ROD - LENGTH TO SUIT, WELDED TO POST

PARLOR FLOOR

WIRE MESH, GROUND ROD, AND 1/4" STEEL ROD ALL CLAMPED TOGETHER TO ENSURE CIRCUIT CONTINUITY

OPTIONAL TECHNIQUE FOR BONDING PARTS
(see notes)

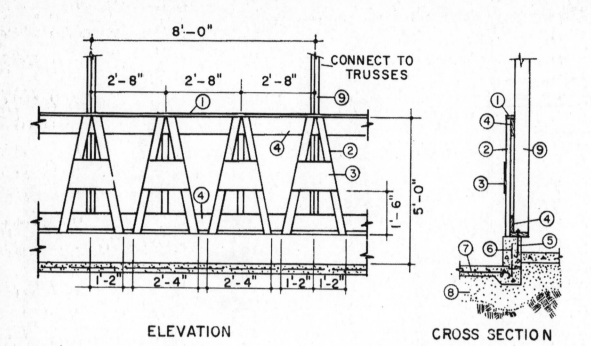

ELEVATION

CROSS SECTION

NOTES

1. 2" X 4" X 16'-0" CAP
2. 2" X 4" X 4'-0" HAY RACK MEMBER
3. 1" X 12" X 1'-9"   "   "   "
4. 2" X 6" X 16'-0"

5. 3/8" ANCHOR BOLT 4'-0" O.C.
6. NO. 3 RE. ROD, 1'-6" LONG, 12" O.C.
7. 6" X 6", 10/10 REINFORCING MESH
8. GRAVEL FILL
9. TWO 2" X 6" X 7'-0" STALL POST

HAY RACK

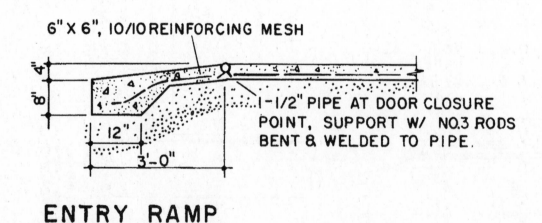

6" X 6", 10/10 REINFORCING MESH

1-1/2" PIPE AT DOOR CLOSURE POINT, SUPPORT W/ NO.3 RODS BENT & WELDED TO PIPE.

ENTRY RAMP

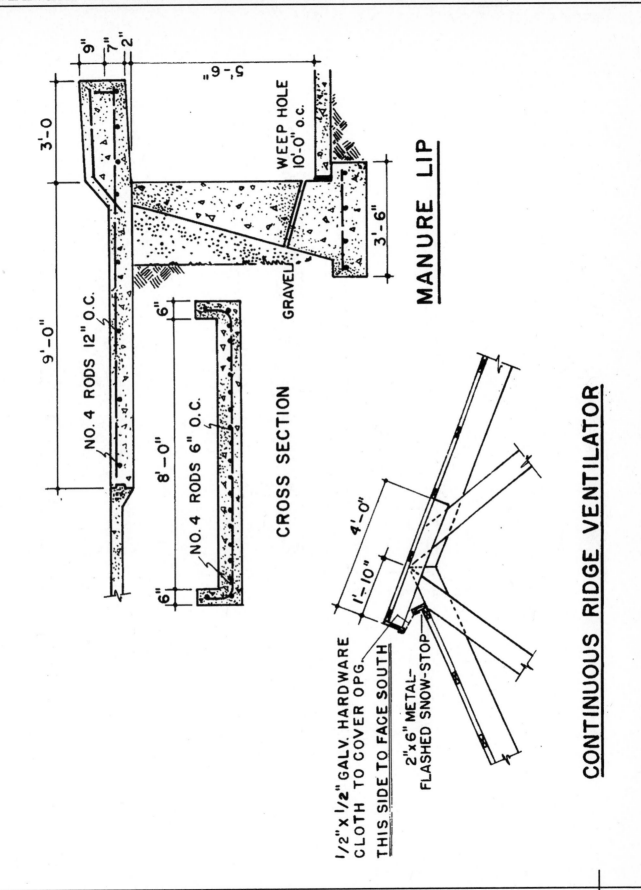

MANURE LIP

CROSS SECTION

CONTINUOUS RIDGE VENTILATOR

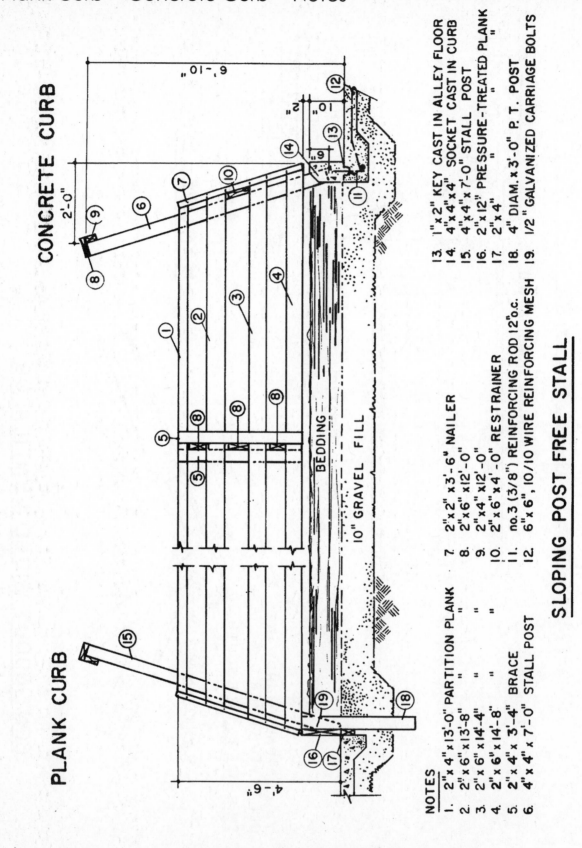

CONCRETE CURB

PLANK CURB

SLOPING POST FREE STALL

NOTES
1. 2"x 4"x13'-0" PARTITION PLANK
2. 2"x 6"x 13'-8" "
3. 2"x 6"x14'-4" "
4. 2"x 6"x14'-8" "
5. 2"x 4"x 3'-4" BRACE
6. 4"x 4"x 7'-0" STALL POST
7. 2"x 2"x3'-6" NAILER
8. 2"x 6"x12'-0"
9. 2"x 4"x12'-0"
10. 2"x 6"x 4'-0" RESTRAINER
11. no.3 (3/8") REINFORCING ROD 12"o.c.
12. 6"x 6", 10/10 WIRE REINFORCING MESH
13. 1"x 2" KEY CAST IN ALLEY FLOOR
14. 4"x 4"x 4" SOCKET CAST IN CURB
15. 4"x 4"x 7'-0" STALL POST
16. 2"x 12" PRESSURE-TREATED PLANK
17. 2"x 4"
18. 4" DIAM. x 3'-0" P.T. POST
19. 1/2" GALVANIZED CARRIAGE BOLTS

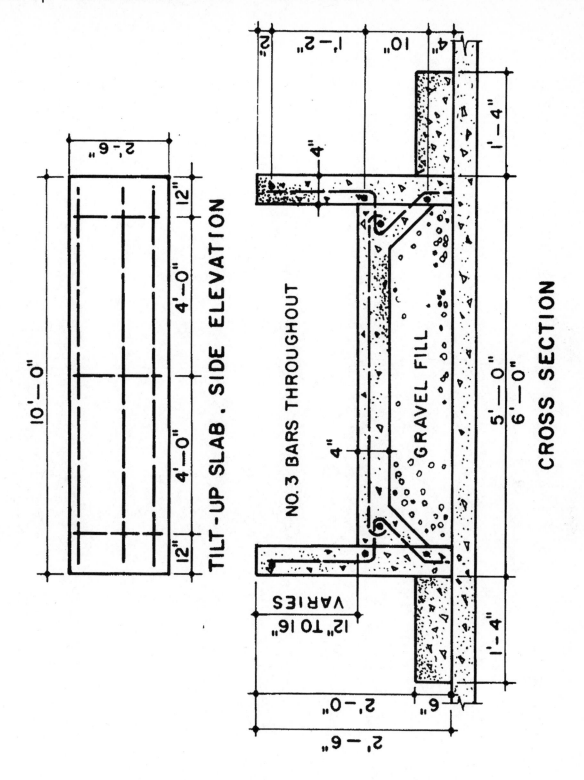

TILT-UP SLAB SIDE ELEVATION

CROSS SECTION

NO. 3 BARS THROUGHOUT

GRAVEL FILL

VARIES
12" TO 16"

The Complete Guide to Building Classic Barns, Fences, Storage Sheds,
Animal Pens, Outbuildings, Greenhouses, Farm Equipment, & Tools

chapter 2

CATTLE BARNS, HUTCHES, PENS, & EQUIPMENT

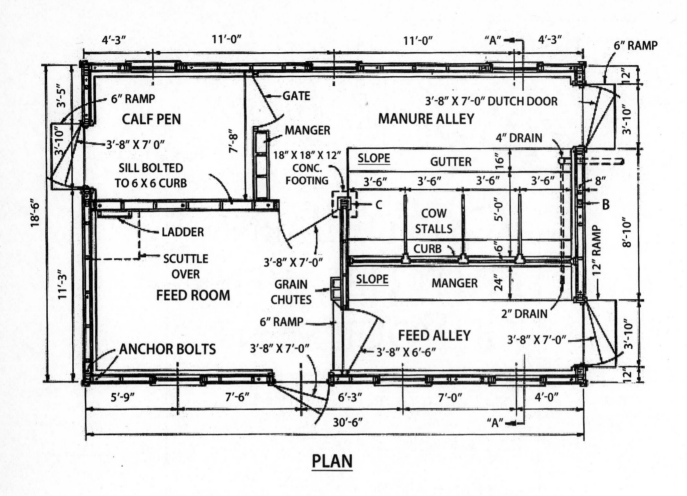

PLAN

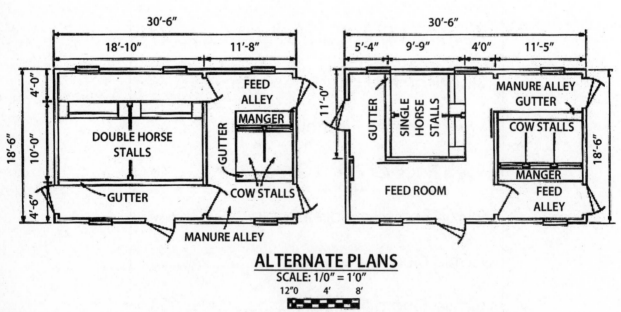

ALTERNATE PLANS
SCALE: 1/0" = 1'0"

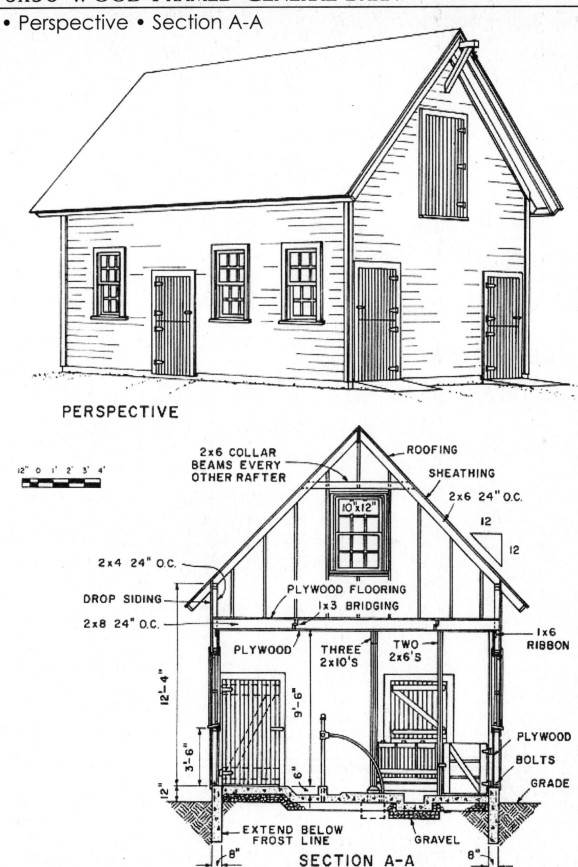

PERSPECTIVE

SECTION A-A

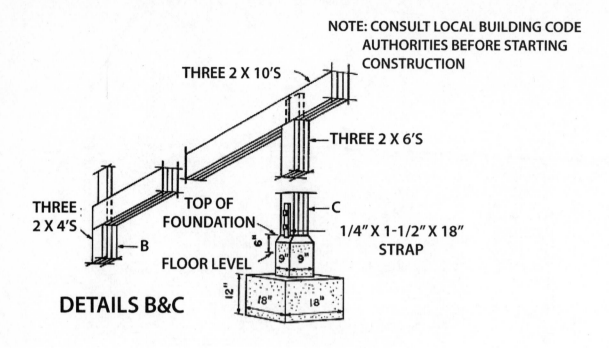

NOTE: CONSULT LOCAL BUILDING CODE AUTHORITIES BEFORE STARTING CONSTRUCTION

THREE 2 X 10'S

THREE 2 X 6'S

THREE 2 X 4'S

B

TOP OF FOUNDATION

C

1/4" X 1-1/2" X 18" STRAP

FLOOR LEVEL

6"

9" 9"

12"

18" 18"

DETAILS B&C

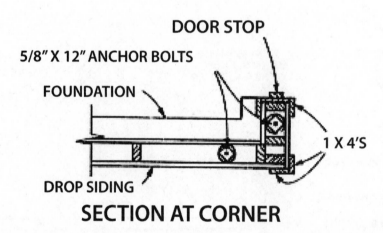

DOOR STOP

5/8" X 12" ANCHOR BOLTS

FOUNDATION

1 X 4'S

DROP SIDING

SECTION AT CORNER

12" 6" 0 1'

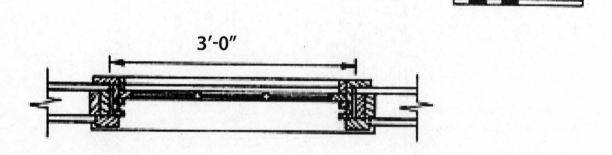

3'-0"

SECTION OF WINDOW

PARTIAL HAYLOFT

• Plan

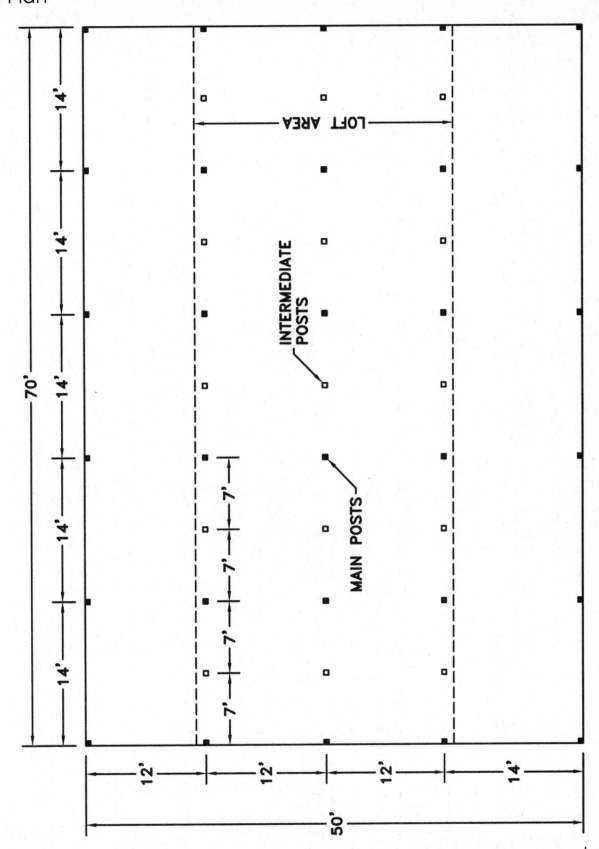

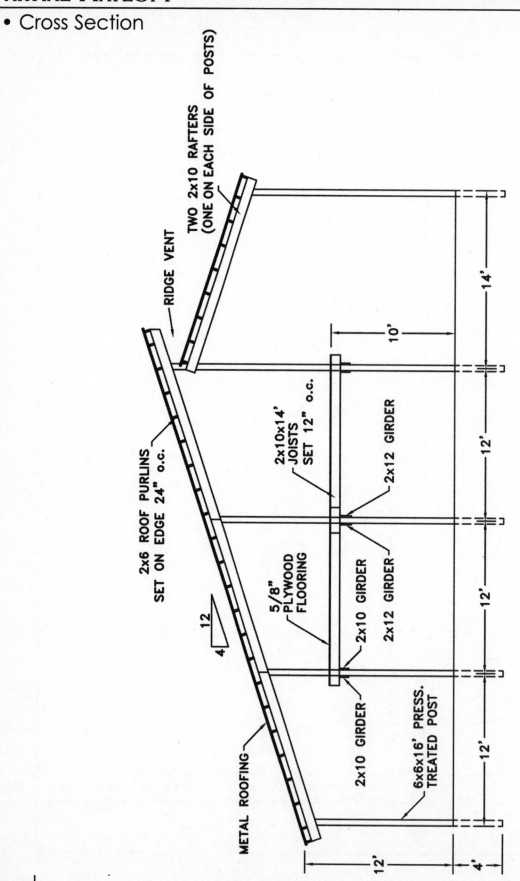

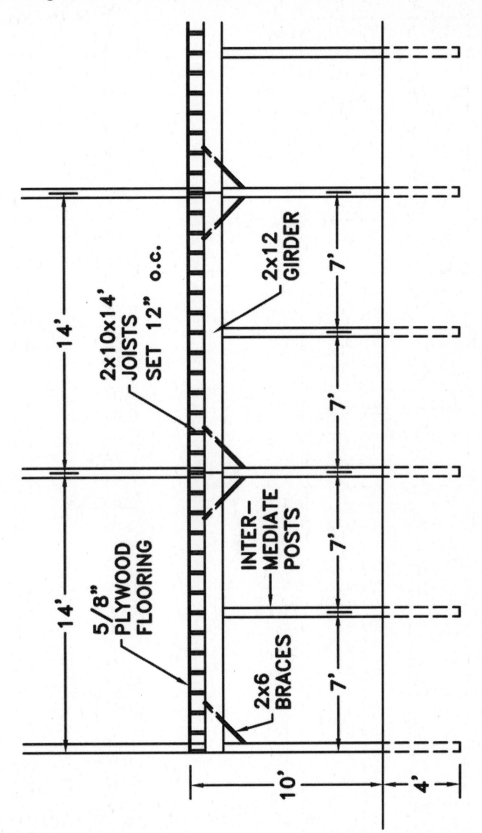

14'

14'

2x10x14'
JOISTS
SET 12" o.c.

5/8"
PLYWOOD
FLOORING

2x12
GIRDER

INTER–
MEDIATE
POSTS

2x6
BRACES

7'

7'

7'

7'

10'

4'

• Top View • Perspective

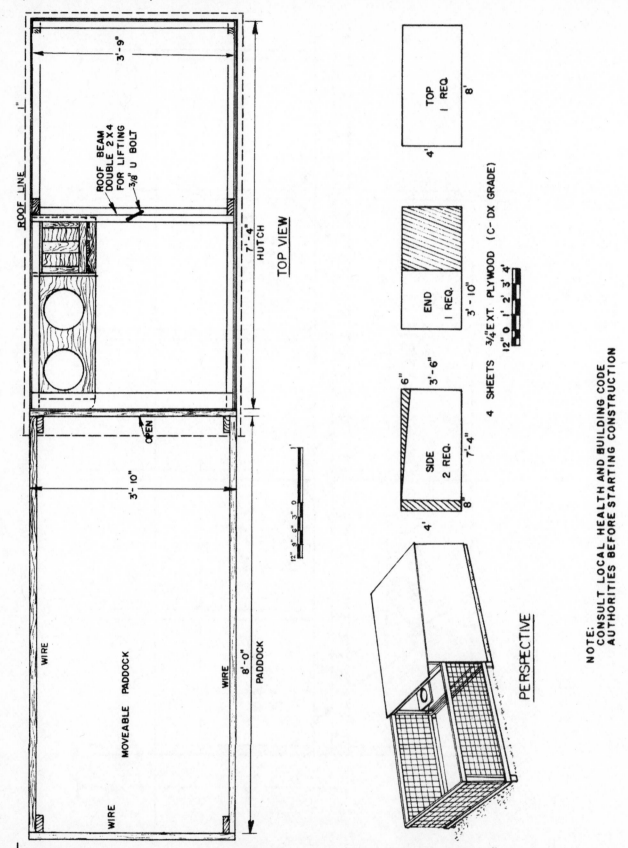

ROOF LINE

3'-9"

1"

ROOF BEAM
DOUBLE 2 X 4
FOR LIFTING

3/8" U BOLT

7'-4"
HUTCH

TOP VIEW

OPEN

3'-10"

WIRE

WIRE

MOVEABLE PADDOCK

8'-0"
PADDOCK

WIRE

TOP
1 REQ.

8'

4'

END
1 REQ.

3'-10"

3/4" EXT. PLYWOOD (C- DX GRADE)

12" 0 1' 2' 3' 4'

4 SHEETS

6"

3'-6"

SIDE
2 REQ.

7'-4"

8'

4'

PERSPECTIVE

12" 9" 6" 3" 0

NOTE:
CONSULT LOCAL HEALTH AND BUILDING CODE
AUTHORITIES BEFORE STARTING CONSTRUCTION

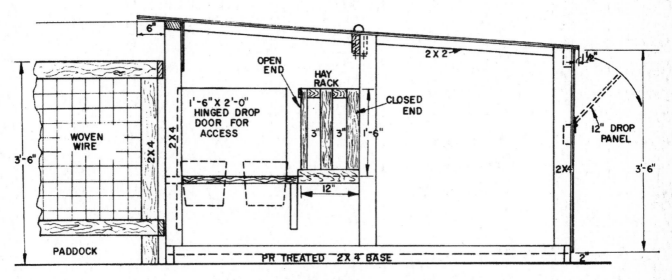

SIDE VIEW
(SIDE COVERING OMITTED)

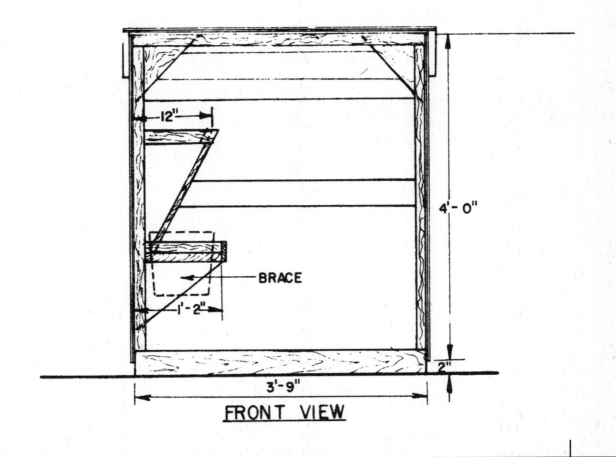

FRONT VIEW

LIST OF MATERIALS
(ALL SOUND QUALITY PLANED HARDWOOD LUMBER HAVING GOOD RESISTANCE TO DECAY)

SIZE*	LENGTH	QUANTITY
3/4" X 2"	19-1/2"	1 PER STALL
3/4" X 2"	21-7/8"	1 PER STALL
3/4" X 2"	4'3"	2 PER STALL
3/4" X 3"	19-1/2"	1 PER STALL
3/4" X 3"	21-7/8"	10 PER STALL
3/4" X 3"	27"	2 PER STALL
3/4" X 3"	4'0"	7 PER STALL
3/4" X 3"	5'0"	1 PER STALL
3/4" X 4"	4'6"**	2 PER STALL
2" X 3"	4'3"	1 PER STALL
2" X 3"	4'0"	1 PER STALL
3/4" X 2"	12'	1 PER 6 STALLS
3/4" X 3"	12'	5 PER 6 STALLS
3/4" X 4"	12'	2 PER 6 STALLS
1/2" PLY OR 3/4" BD	10" X 19"	1 PER STALL
1/2" PLY.	12" X 16" SLOPED	2 PER STALL

SIZE*	LENGTH	QUANITY
NAILS:	1-1/2" (COMMON OR HARDENED)	1/10 LB. PER STALL
	2" (COMMON OR HARDENED)	1/3 LB. PER STALL
BOLTS:	3/8" X 4" MACH.	2 PER STALL
WASHERS:	3/8" FLAT	4 PER STALL

* ALL WOOD SIZES "ACTUAL"

** 4'-6" FOR VEAL CALVES TO 300#, ONLY 4'-0" FOR CALVES TO 200#

ISOMETRIC

• Front View

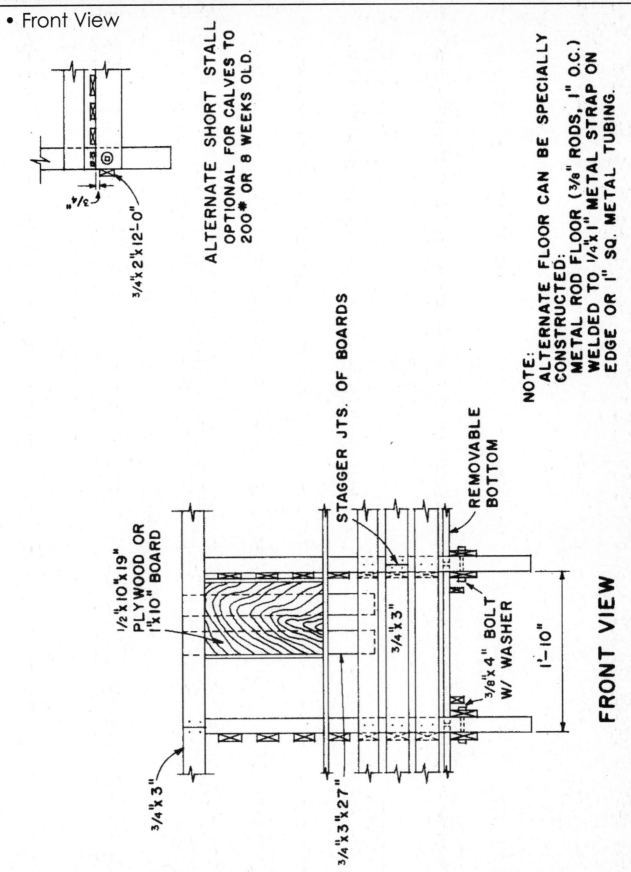

3/4" x 2" x 12'-0"

ALTERNATE SHORT STALL
OPTIONAL FOR CALVES TO
200# OR 8 WEEKS OLD.

NOTE:
ALTERNATE FLOOR CAN BE SPECIALLY
CONSTRUCTED:
METAL ROD FLOOR (3/8" RODS, 1" O.C.)
WELDED TO 1/4" x 1" METAL STRAP ON
EDGE OR 1" SQ. METAL TUBING.

STAGGER JTS. OF BOARDS

REMOVABLE BOTTOM

1/2" x 10" x 19" PLYWOOD OR 1" x 10" BOARD

3/4" x 3"

3/4" x 3"

3/4" x 3" x 27"

3/8" x 4" BOLT W/ WASHER

1'-10"

FRONT VIEW

• Side View

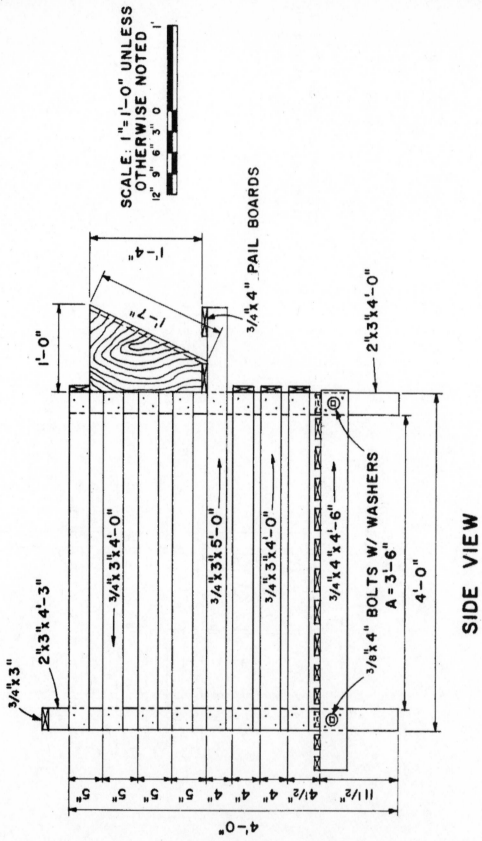

SCALE: 1"=1'-0" UNLESS OTHERWISE NOTED

3/4" x 4" PAIL BOARDS

1'-4"

1'-7"

1'-0"

2"x3"x4'-0"

2"x3"x4'-3"

3/4" x 3"

3/4" x 3" x 4'-0"

3/4" x 3" x 5'-0"

3/4" x 3" x 4'-0"

3/4" x 4" x 4'-6"

3/8" x 4" BOLTS W/ WASHERS
A = 3'-6"

4'-0"

5" 5" 5" 5" 4" 4" 4" 4½" 4½" 11½"

4'-0"

SIDE VIEW

Elevated Wood Calf Stall

• Removeable Bottom–Top View

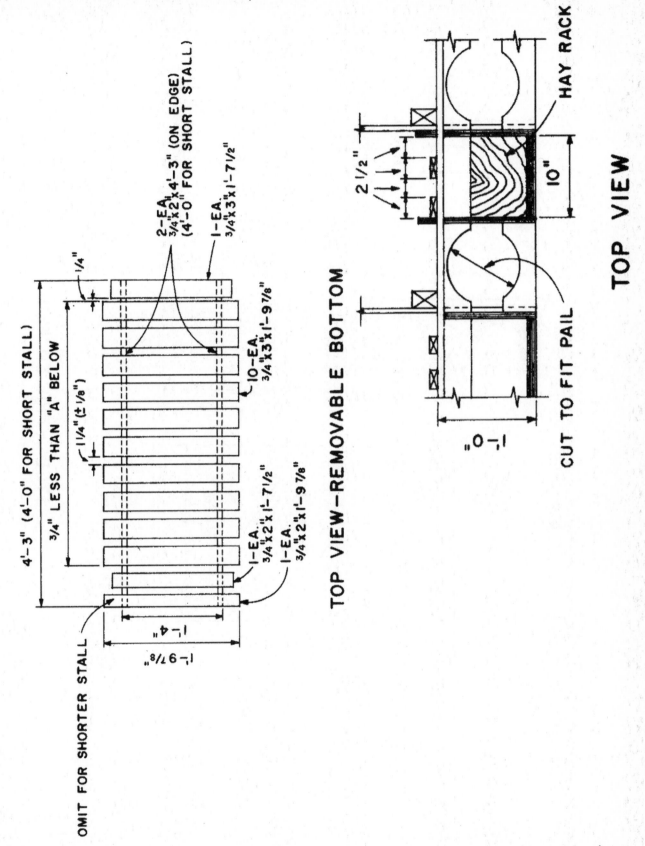

TOP VIEW–REMOVABLE BOTTOM

TOP VIEW

2–EA.
3/4"x2"x4'–3" (ON EDGE)
(4'–0" FOR SHORT STALL)

1–EA.
3/4"x3"x1'–7 1/2"

10–EA.
3/4"x3"x1'–9 7/8"

1–EA.
3/4"x2"x1'–7 1/2"

1–EA.
3/4"x2"x1'–9 7/8"

4'–3" (4'–0" FOR SHORT STALL)

3/4" LESS THAN "A" BELOW

1 1/4" (± 1/8")

1/4"

1'–4"

1'–9 7/8"

OMIT FOR SHORTER STALL

HAY RACK

2 1/2"

10"

1'–0"

CUT TO FIT PAIL

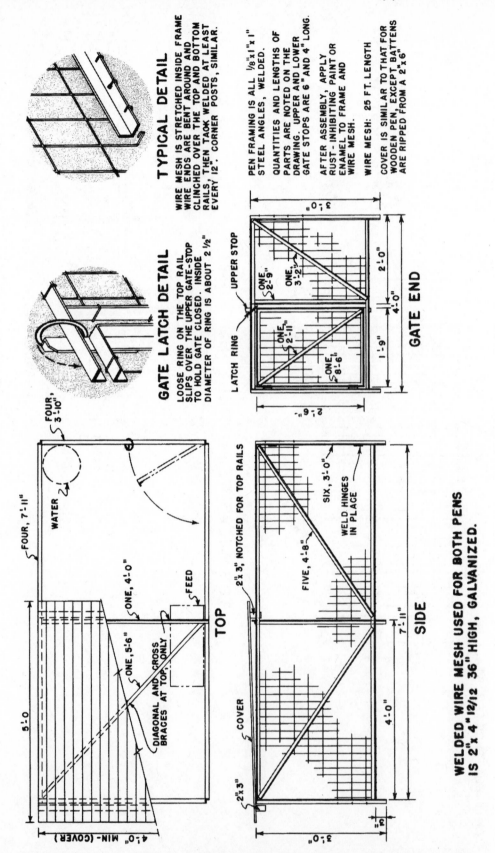

TYPICAL DETAIL

WIRE MESH IS STRETCHED INSIDE FRAME WIRE ENDS ARE BENT AROUND AND CLINCHED OVER THE TOP AND BOTTOM RAILS, THEN TACK WELDED AT LEAST EVERY 12". CORNER POSTS, SIMILAR.

PEN FRAMING IS ALL 1/8"x1"x1" STEEL ANGLES, WELDED.

QUANTITIES AND LENGTHS OF PARTS ARE NOTED ON THE DRAWING. UPPER AND LOWER GATE STOPS ARE 6" AND 4" LONG.

AFTER ASSEMBLY, APPLY RUST-INHIBITING PAINT OR ENAMEL TO FRAME AND WIRE MESH.

WIRE MESH: 25 FT. LENGTH

COVER IS SIMILAR TO THAT FOR WOODEN PEN, EXCEPT BATTENS ARE RIPPED FROM A 2"x6"

GATE LATCH DETAIL

LOOSE RING ON THE TOP RAIL SLIPS OVER THE UPPER GATE-STOP TO HOLD GATE CLOSED. INSIDE DIAMETER OF RING IS ABOUT 2 1/2"

GATE END

UPPER STOP
LATCH RING
ONE, 2'-9"
ONE, 3'-2"
ONE, 2'-11"
ONE, 8'-6"
3'-0"
2'-0"
4'-0"
1'-9"
2'-6"

TOP

FOUR, 7'-11"
FOUR, 3'-10"
WATER
ONE, 4'-0"
FEED
ONE, 5'-6"
DIAGONAL AND CROSS BRACES AT TOP ONLY
5'-0"
4'-0" MIN-(COVER)

SIDE

2"x3", NOTCHED FOR TOP RAILS
SIX, 3'-0"
WELD HINGES IN PLACE
FIVE, 4'-8"
COVER
2"x3"
7'-11"
4'-0"
3'-0"

WELDED WIRE MESH USED FOR BOTH PENS IS 2"x4" 12/12 36" HIGH, GALVANIZED.

STEEL-FRAMED PEN

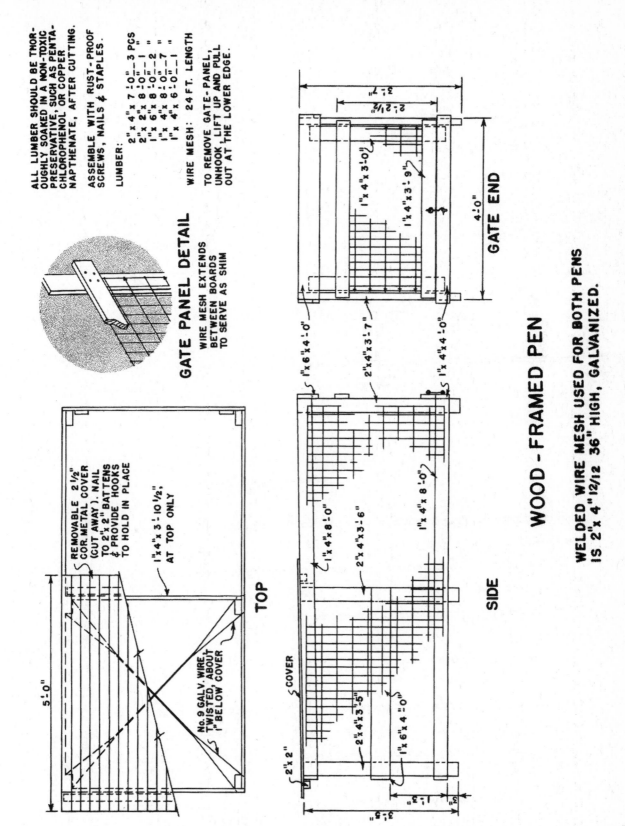

ALL LUMBER SHOULD BE THOR-OUGHLY SOAKED IN A NON-TOXIC PRESERVATIVE, SUCH AS PENTA-CHLOROPHENOL, OR COPPER NAPTHENATE, AFTER CUTTING.

ASSEMBLE WITH RUST-PROOF SCREWS, NAILS & STAPLES.

LUMBER:

2"x 4"x 7'-0"—3 PCS
2"x 4"x 8'-0"—2
1"x 6"x 8'-0"—2
1"x 4"x 8'-0"—7
1"x 4"x 6'-0"—11

WIRE MESH: 24 FT. LENGTH

TO REMOVE GATE-PANEL, UNHOOK, LIFT UP AND PULL OUT AT THE LOWER EDGE.

GATE PANEL DETAIL

WIRE MESH EXTENDS BETWEEN BOARDS TO SERVE AS SHIM

GATE END

1"x 4"x 3'-0"
1"x 4"x 3'-9"

1"x 6"x 4'-0"
2"x 4"x 3'-7"
1"x 4"x 4'-0"

WOOD-FRAMED PEN

WELDED WIRE MESH USED FOR BOTH PENS IS 2"x 4" 12/12 36" HIGH, GALVANIZED.

TOP

REMOVABLE 2½" COR. METAL COVER (CUT AWAY). NAIL TO 2"x 2" BATTENS & PROVIDE HOOKS TO HOLD IN PLACE

1"x 4"x 3'-10½", AT TOP ONLY

5'-0"

No. 9 GALV. WIRE TWISTED, ABOUT 1" BELOW COVER

SIDE

1"x 4"x 8'-0"
2"x 4"x 3'-6"

COVER

2"x 2"
2"x 4"x 3'-5"
1"x 6"x 4'-0"

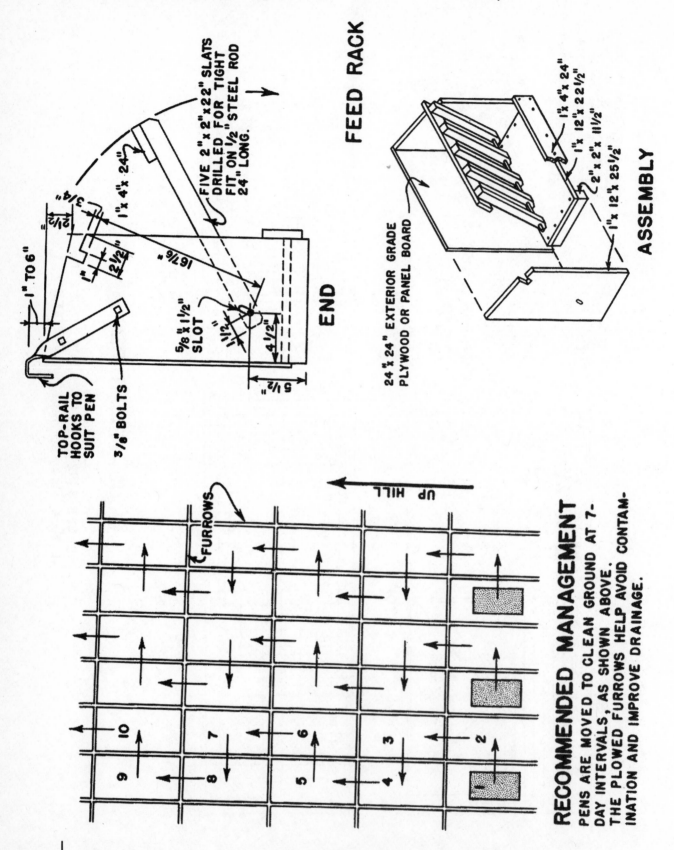

FEED RACK

END

FIVE 2"x 2"x 22" SLATS DRILLED FOR TIGHT FIT ON 1/2" STEEL ROD 24" LONG.

1"x 4"x 24"

1" TO 6"

3/4"

2 1/2"

1"

2 1/2"

16 7/8"

5/8"x 1 1/2" SLOT

1 1/2"

4 1/2"

5 1/2"

TOP-RAIL HOOKS TO SUIT PEN

3/8" BOLTS

ASSEMBLY

1"x 4"x 24"

1"x 12"x 22 1/2"

2"x 2"x 11 1/2"

1"x 12"x 25 1/2"

24"x 24" EXTERIOR GRADE PLYWOOD OR PANEL BOARD

RECOMMENDED MANAGEMENT

PENS ARE MOVED TO CLEAN GROUND AT 7-DAY INTERVALS, AS SHOWN ABOVE. THE PLOWED FURROWS HELP AVOID CONTAM-INATION AND IMPROVE DRAINAGE.

UP HILL

FURROWS

1 2 3 4 5 6 7 8 9 10

9X4X4 Cattle Breeding Rack

• List of Materials • Perspective & End Views

LIST OF MATERIALS

(2) 2" X 12" X 10'-0" BASES
(2) 2" X 12" X 10'-0" UPRIGHTS
(1) 2" X 8" X 6'-0" UPRIGHT
(1) 2" X 4" X 12'-0" BOTTOM TIE
(2) 2" X 6" X 10'-0" DIAGONAL BRACES
(2) 2" X 4" X 12'-0" EXTENSION BARS
(2) 2" X 12" X 10'-0" FOOT RESTS
(1) 2" X 4" X 8'-0" END TIE
(1) 2" X 4" X 10'-0" YOKE
(1) 1" X 4" X 6'-0" YOKE
(1) 1" X 3" X 8'-0" FOOT GUARD
30 LIN. FT. 1" X 2" CLEATS
6 LBS. 20 D NAILS
2 LBS. 10 D NAILS
(1) 3/4" EYE BOLT 4" LONG 1-1/2" EYE

NOTE:

THE YOKE MAY BE MOVED FORWARD OR BACK TO SUIT SIZE OF ANIMAL.

ALL WOOD SHOULD BE TREATED WITH PRESERVATIVE.

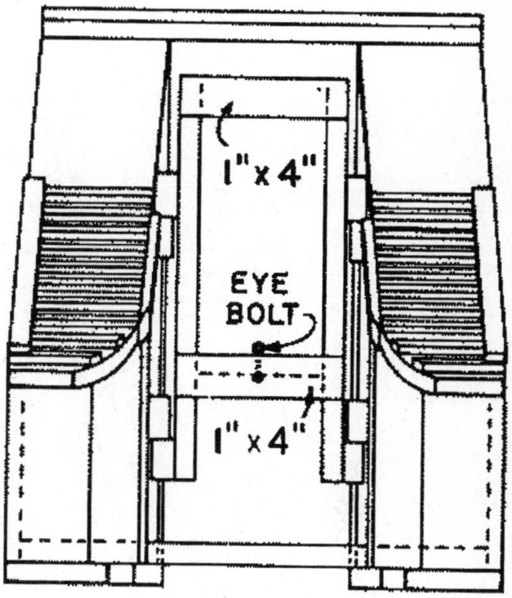

END VIEW

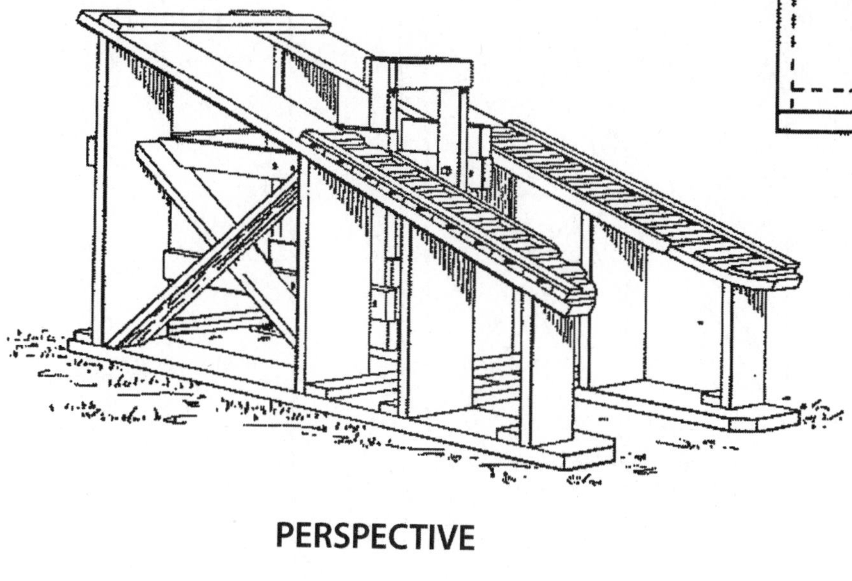

PERSPECTIVE

9X4X4 CATTLE BREEDING RACK

• Elevation

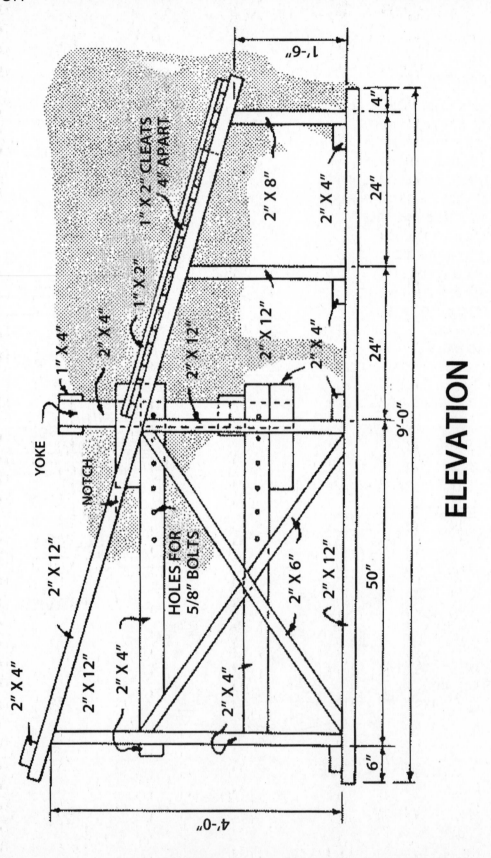

50

THE COMPLETE GUIDE TO BUILDING CLASSIC BARNS, FENCES, STORAGE SHEDS, ANIMAL PENS, OUTBUILDINGS, GREENHOUSES, FARM EQUIPMENT, & TOOLS

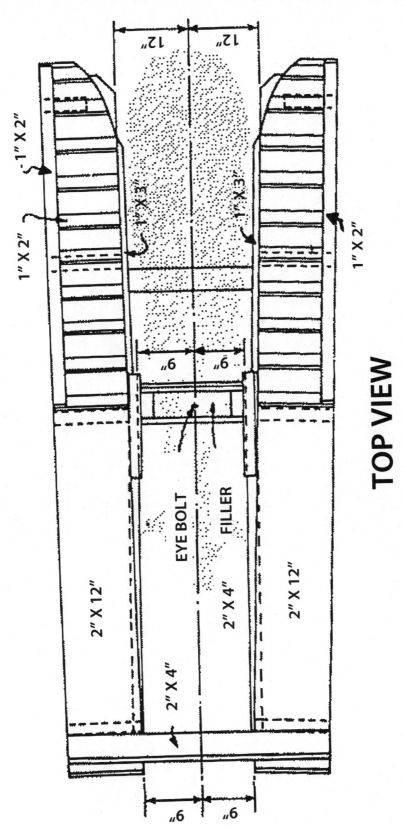

TOP VIEW

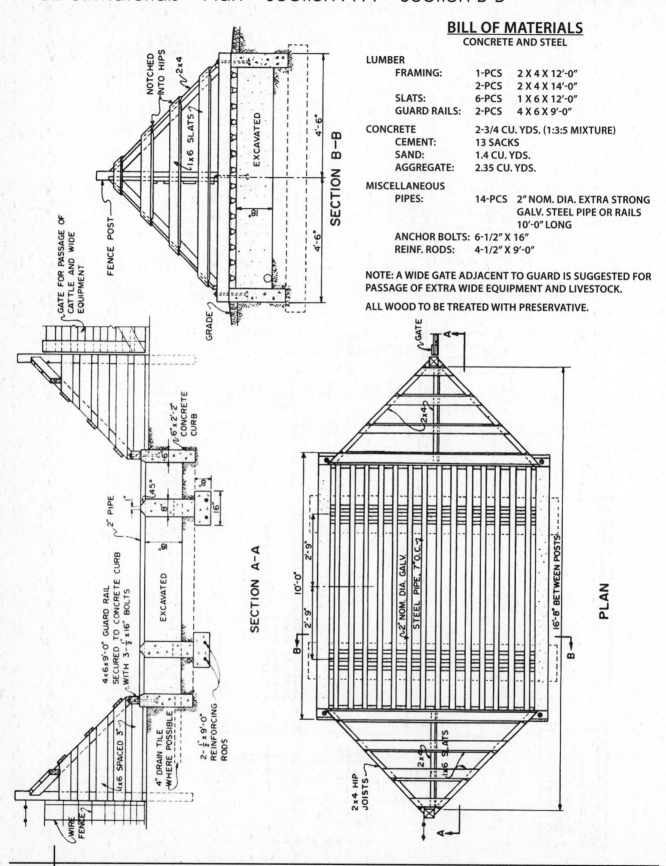

BILL OF MATERIALS
CONCRETE AND STEEL

LUMBER
FRAMING:	1-PCS	2 X 4 X 12'-0"
	2-PCS	2 X 4 X 14'-0"
SLATS:	6-PCS	1 X 6 X 12'-0"
GUARD RAILS:	2-PCS	4 X 6 X 9'-0"

CONCRETE   2-3/4 CU. YDS. (1:3:5 MIXTURE)
CEMENT:	13 SACKS
SAND:	1.4 CU. YDS.
AGGREGATE:	2.35 CU. YDS.

MISCELLANEOUS
PIPES:	14-PCS	2" NOM. DIA. EXTRA STRONG GALV. STEEL PIPE OR RAILS 10'-0" LONG
ANCHOR BOLTS:	6-1/2" X 16"	
REINF. RODS:	4-1/2" X 9'-0"	

NOTE: A WIDE GATE ADJACENT TO GUARD IS SUGGESTED FOR PASSAGE OF EXTRA WIDE EQUIPMENT AND LIVESTOCK.

ALL WOOD TO BE TREATED WITH PRESERVATIVE.

THE COMPLETE GUIDE TO BUILDING CLASSIC BARNS, FENCES, STORAGE SHEDS, ANIMAL PENS, OUTBUILDINGS, GREENHOUSES, FARM EQUIPMENT, & TOOLS

BILLS OF MATERIAL

CONCRETE AND WOOD

LUMBER
FENCE POSTS: 2-PCS 4x4x7'-0"
FRAMING: 3-PCS 2x12x10'-0"
7-PCS 2x12x8'-0"
13-PCS 2x4x10'-0"
2-PCS 2x4x9'-0"
3-PCS 2x4x8'-0"
6-PCS 2x4x6'-0"
SLATS: 9-PCS 1x6x8'-0"
GUARD RAILS: 2-PCS 4x6x8'-0"
CONCRETE 2½ CU. YDS.
(1:3:5 MIXTURE)
CEMENT: 12 SACKS
SAND: 1.25 CU. YDS.
AGGREGATE: 2.13 CU. YDS.
MISCELLANEOUS
ANCHOR BOLTS: FOUR ½"x12" GALV.
REINF. RODS: FOUR ½" x 9'-9"
TWO ½" x 8'-9"

WOOD

LUMBER
FENCE POSTS: 2-PCS 4x4x7'-0"
FRAMING: 2-PCS 12x12x9'-0"
2-PCS 4x4x10'-0"
3-PCS 2x12x10'-0"
9-PCS 2x12x8'-0"
11-PCS 2x4x10'-0"
3-PCS 2x4x8'-0"
6-PCS 2x4x6'-0"
SLATS: 9-PCS 1x6x8'-0"
GUARD RAILS: 2-PCS 4x6x8'-0"
MISCELLANEOUS
ANCHOR BOLTS: FOUR ½"x12" GALV.

NOTE:

USE ROUGH-SAWN LUMBER.
ALL WOOD IS TO BE PRESSURE TREATED
WITH PRESERVATIVE.

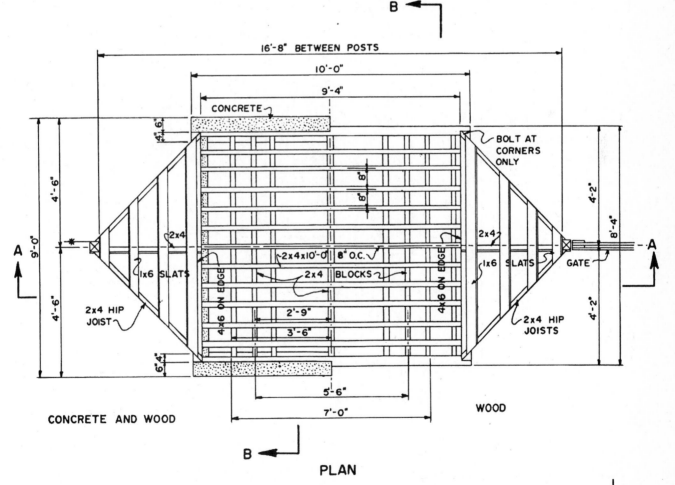

PLAN

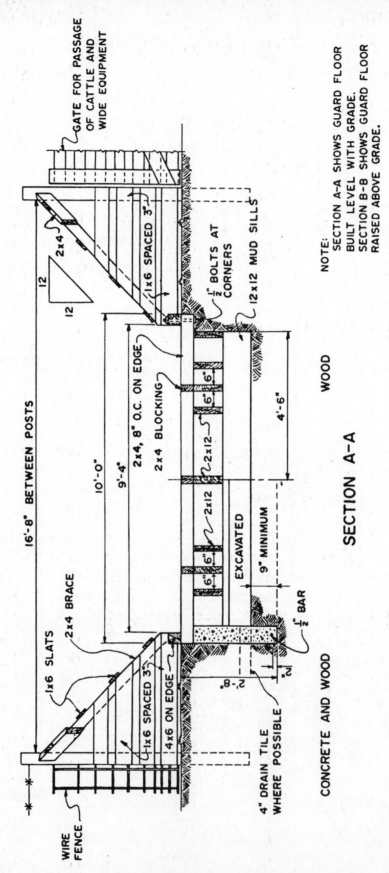

GATE FOR PASSAGE OF CATTLE AND WIDE EQUIPMENT

2x4

12
12

1x6 SPACED 3"

½" BOLTS AT CORNERS

12x12 MUD SILLS

NOTE:
SECTION A-A SHOWS GUARD FLOOR BUILT LEVEL WITH GRADE.
SECTION B-B SHOWS GUARD FLOOR RAISED ABOVE GRADE.

16'-8" BETWEEN POSTS

10'-0"

9'-4"

2x4, 8" O.C. ON EDGE

2x4 BLOCKING

6" 6"

6" 6"   2x12

2x12

6" 6"

EXCAVATED

9" MINIMUM

4'-6"

WOOD

½" BAR

2"

2'-8"

2x4 BRACE

1x6 SLATS

1x6 SPACED 3"

4x6 ON EDGE

4" DRAIN TILE WHERE POSSIBLE

CONCRETE AND WOOD

WIRE FENCE

SECTION A-A

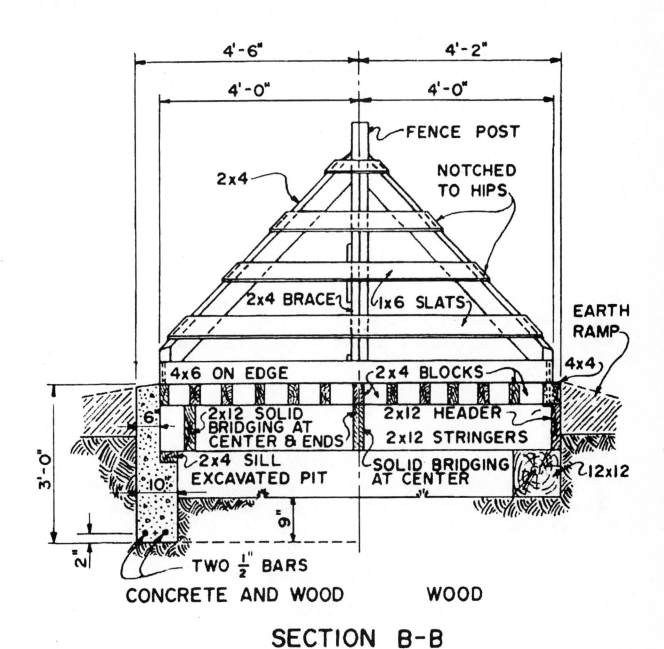

SECTION B-B

Mark Johnson
Sales manager, Green Design and Construction LLC
info@greendesignandconstruction.com
www.greendesignandconstruction.com

Mark Johnson started his building career in 1973 by building pole barns, and in 1976, he built his first home from a timber frame barn he salvaged, moved, and rebuilt. He now buys, restores, and builds pioneer farm log cabins and timber frame barns into homes and barns.

Johnson said the three most important design considerations when selecting a building plan are functionality, aesthetics, and cost. Before building a new structure, he recommends homeowners perform a comprehensive wants/needs analysis for the building's function. He said to estimate the building's basic cost in a way that puts a dollar amount on the elective options so builders can prioritize each option according to whether it fits within the budget. He also explained the size of the barn or shed will dictate the amount of time and skill, as well as the types of tools, required.

"Every job takes less time with some help, and if the help has experience it's a plus," he said.

Johnson said floating slabs are the most economical flooring option if a floor surface is necessary, but in general, barns and sheds do not require very complex foundations. Floating slabs can resist frost heave, but if the building does not require a floor, post and beam foundations are the most economical option. He suggests using steel siding and roofing on buildings because they not only provide a maintenance-free exterior finish, but they also replace the need for sheathing on a building.

One thing he has noticed from his experience building barns is many people prefer the look of the traditional barn. He recommends salvaging an existing barn or home as a great way to achieve the look of a traditional building.

"There are plenty [of barns] available in the Midwest, and most owners would be glad to see them salvaged and rebuilt instead of bulldozed and burned."

chapter 3

Horse Barn & Trailer

24X36 THREE-STALL HORSE BARN

• Plan • Isometric View

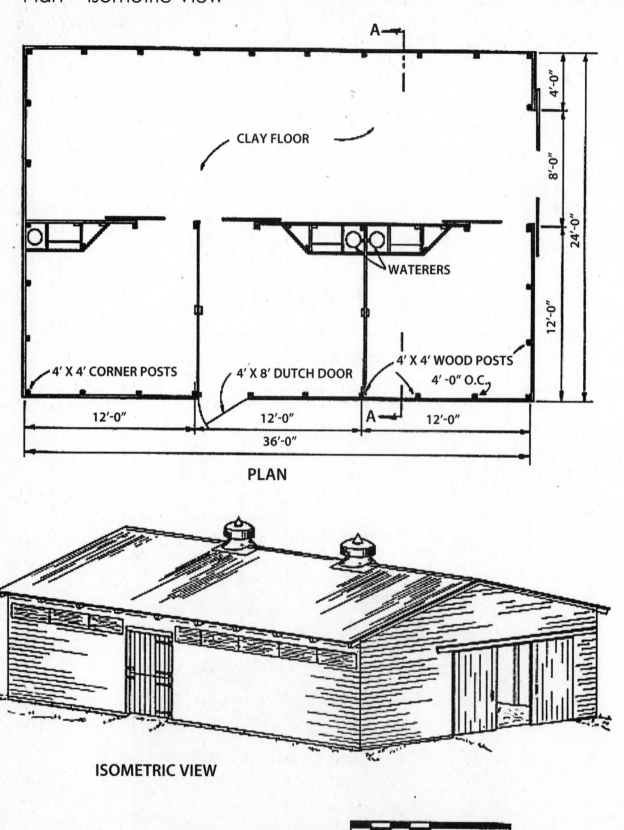

A

CLAY FLOOR

WATERERS

4' X 4' CORNER POSTS

4' X 8' DUTCH DOOR

4' X 4' WOOD POSTS
4' -0" O.C.

A

4'-0"

8'-0"

24'-0"

12'-0"

12'-0"   12'-0"   12'-0"

36'-0"

PLAN

ISOMETRIC VIEW

THE COMPLETE GUIDE TO BUILDING CLASSIC BARNS, FENCES, STORAGE SHEDS, ANIMAL PENS, OUTBUILDINGS, GREENHOUSES, FARM EQUIPMENT, & TOOLS

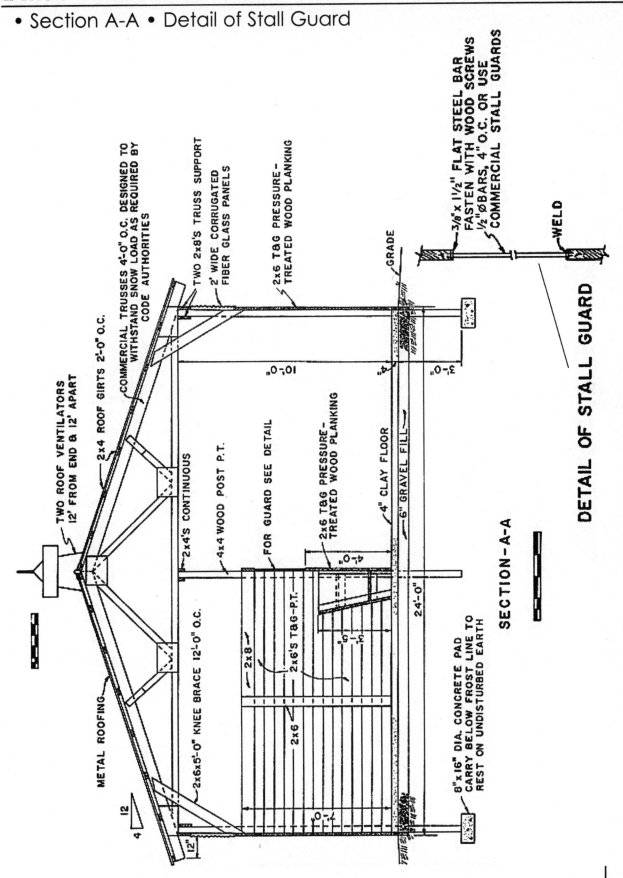

TWO ROOF VENTILATORS 12' FROM END & 12' APART

COMMERCIAL TRUSSES 4'-0" O.C. DESIGNED TO WITHSTAND SNOW LOAD AS REQUIRED BY CODE AUTHORITIES

2x4 ROOF GIRTS 2'-0" O.C.

TWO 2x8'S TRUSS SUPPORT

2' WIDE CORRUGATED FIBER GLASS PANELS

2x6 T&G PRESSURE-TREATED WOOD PLANKING

GRADE

METAL ROOFING

2x6x5'-0" KNEE BRACE 12'-0" O.C.

2x4'S CONTINUOUS

4x4 WOOD POST P.T.

FOR GUARD SEE DETAIL

2x6 T&G PRESSURE-TREATED WOOD PLANKING

4" CLAY FLOOR

6" GRAVEL FILL

2x8

2x6'S T&G-P.T.

2x6

8"x16" DIA. CONCRETE PAD CARRY BELOW FROST LINE TO REST ON UNDISTURBED EARTH

10'-0"

3'-0"

4'-0"

24'-0"

12

4

12"

SECTION-A-A

3/8" x 1 1/2" FLAT STEEL BAR FASTEN WITH WOOD SCREWS 1/2" ØBARS, 4" O.C. OR USE COMMERCIAL STALL GUARDS

WELD

DETAIL OF STALL GUARD

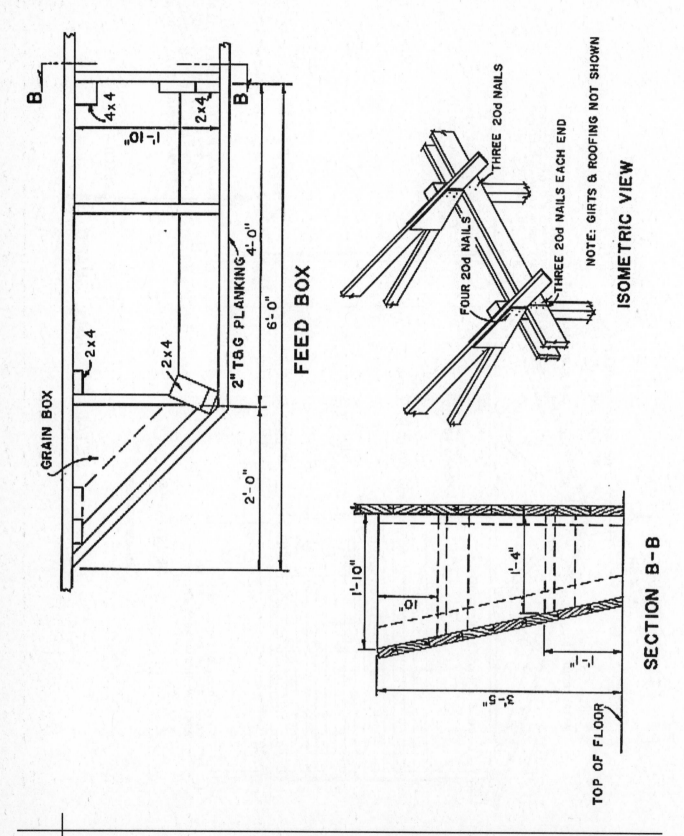

FEED BOX

ISOMETRIC VIEW

NOTE: GIRTS & ROOFING NOT SHOWN

THREE 20d NAILS

THREE 20d NAILS EACH END

FOUR 20d NAILS

GRAIN BOX

4 x 4

2 x 4

2 x 4

2 x 4

2" T&G PLANKING

1'-10"

4'-0"

6'-0"

2'-0"

SECTION B-B

TOP OF FLOOR

1'-10"

10"

1'-4"

1'-1"

3'-5"

• Side Elevation of Trailer

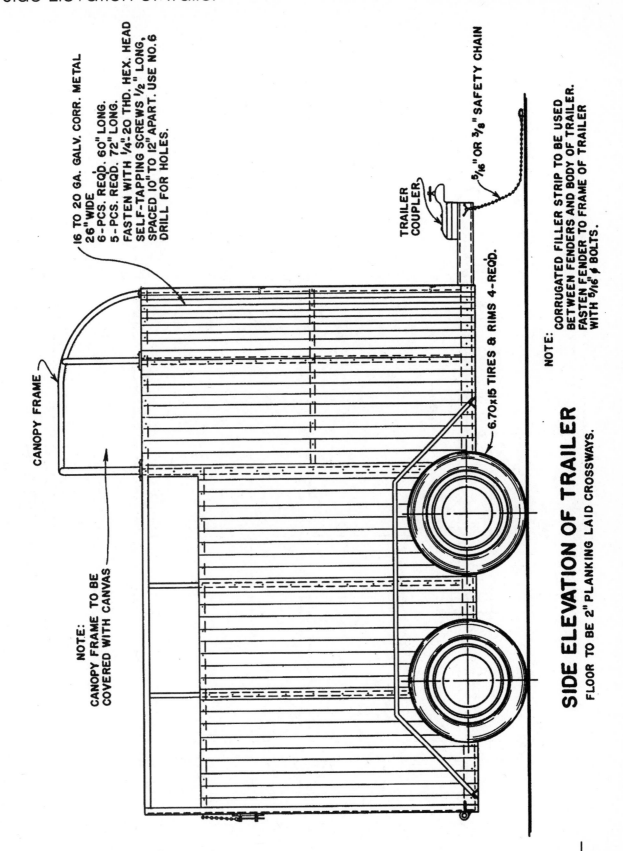

16 TO 20 GA. GALV. CORR. METAL 26" WIDE
6 - PCS. REQ'D. 60" LONG.
5 - PCS. REQ'D. 72" LONG.
FASTEN WITH ¼"-20 THD. HEX. HEAD SELF-TAPPING SCREWS ½" LONG, SPACED 10" TO 12" APART. USE NO.6 DRILL FOR HOLES.

TRAILER COUPLER

⁵⁄₁₆" OR ⅜" SAFETY CHAIN

NOTE:
CORRUGATED FILLER STRIP TO BE USED BETWEEN FENDERS AND BODY OF TRAILER. FASTEN FENDER TO FRAME OF TRAILER WITH ⁵⁄₁₆" ø BOLTS.

6.70x15 TIRES & RIMS 4 - REQ'D.

CANOPY FRAME

NOTE:
CANOPY FRAME TO BE COVERED WITH CANVAS

SIDE ELEVATION OF TRAILER
FLOOR TO BE 2" PLANKING LAID CROSSWAYS.

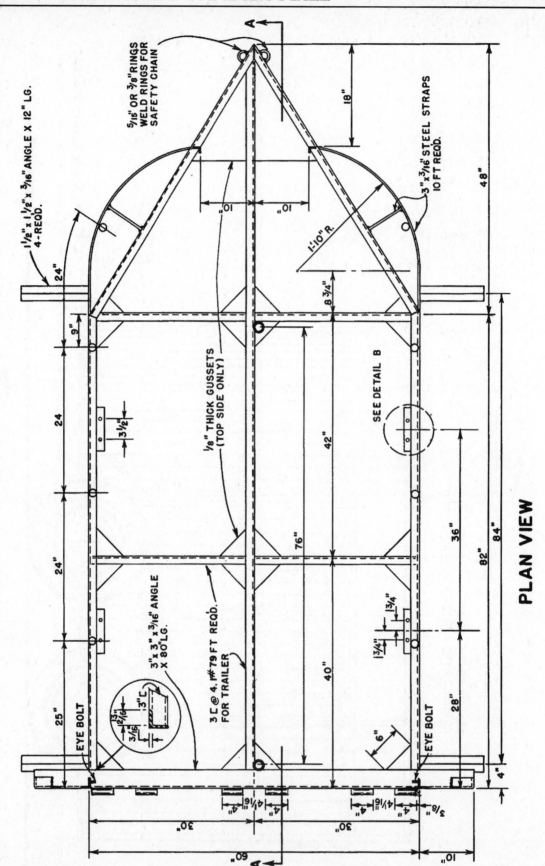

• Side View of Frame

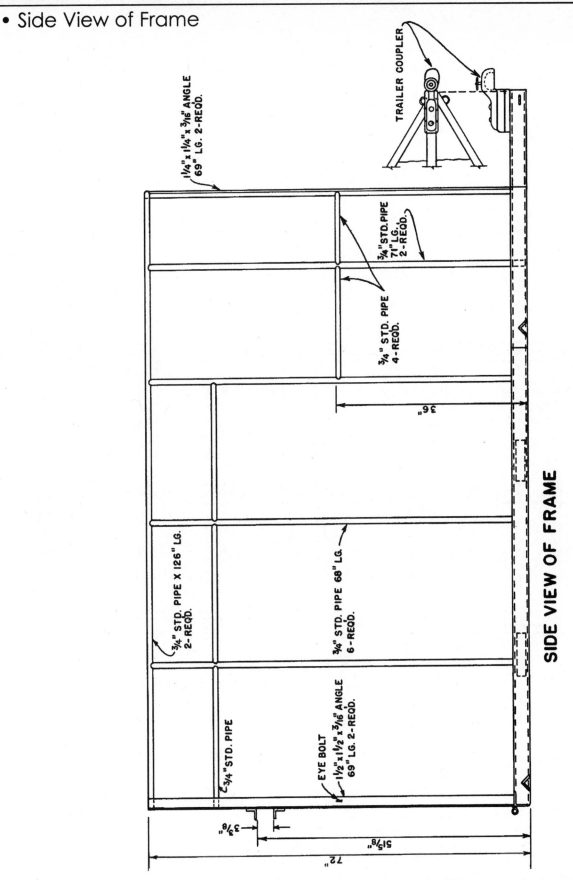

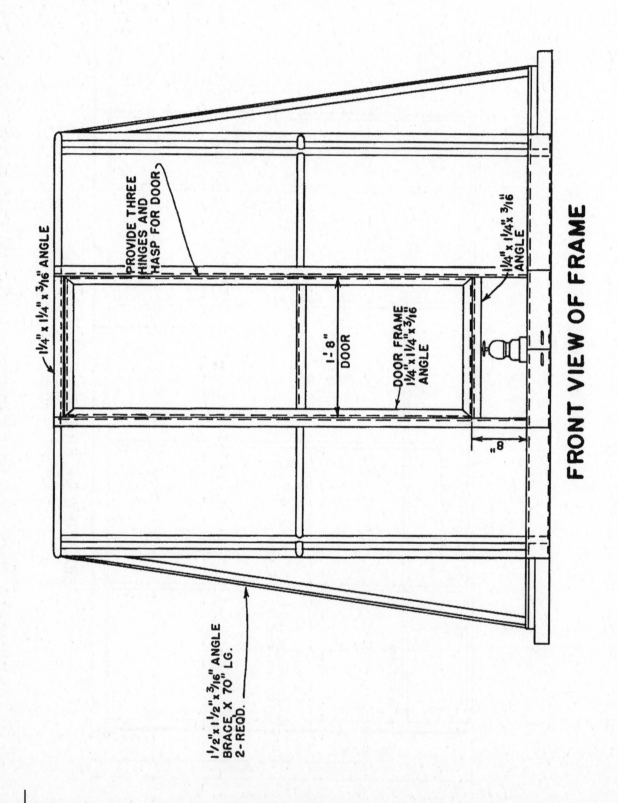

PROVIDE THREE HINGES AND HASP FOR DOOR

1¼" x 1¼" x 3/16" ANGLE

1¼" x 1¼" x 3/16" ANGLE

DOOR FRAME 1¼" x 1¼" x 3/16" ANGLE

1'-8" DOOR

8"

1½" x 1½" x 3/16" ANGLE BRACE X 70" LG. 2-REQD.

FRONT VIEW OF FRAME

• Canopy Frame

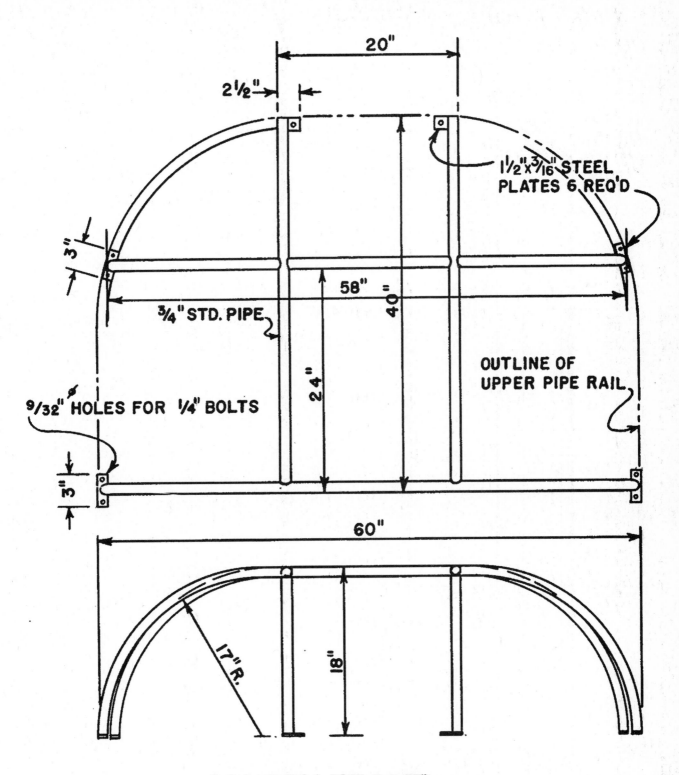

CANOPY FRAME 1-REQ'D

• Fender

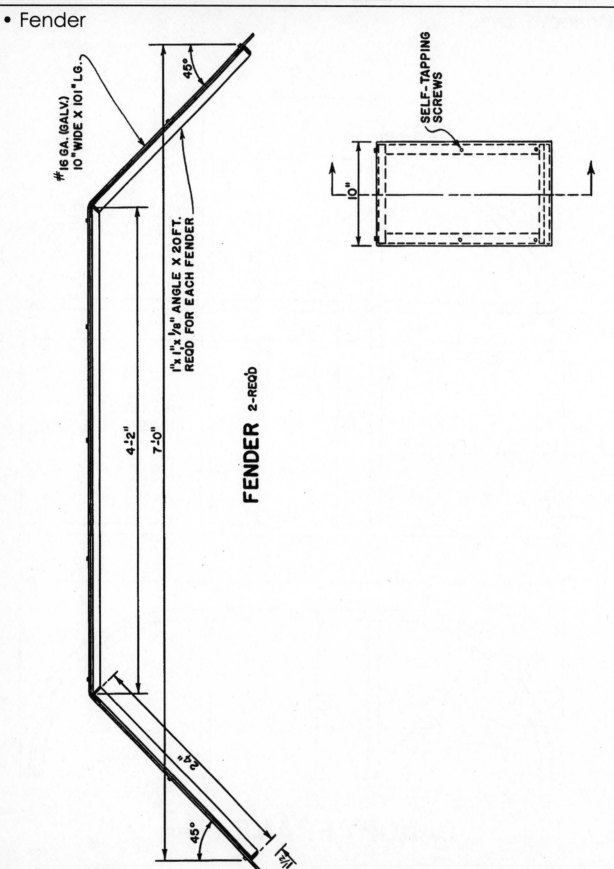

#16 GA. (GALV.) 10" WIDE X 101" LG.

45°

1" x 1" x ⅛" ANGLE X 20 FT. REQ'D FOR EACH FENDER

4'-2"

7'-0"

24"

45°

½"

FENDER 2-REQ'D

SELF-TAPPING SCREWS

10"

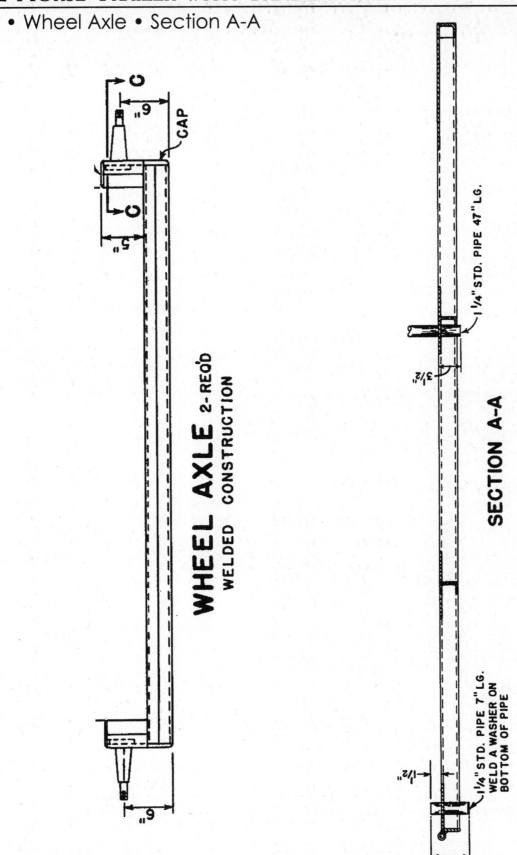

WHEEL AXLE 2- REQ'D
WELDED CONSTRUCTION

9"

C

CAP

5"

9"

1 ¼" STD. PIPE 47" LG.

3½"

SECTION A-A

1 ¼" STD. PIPE 7" LG.
WELD A WASHER ON
BOTTOM OF PIPE

1½"

7"

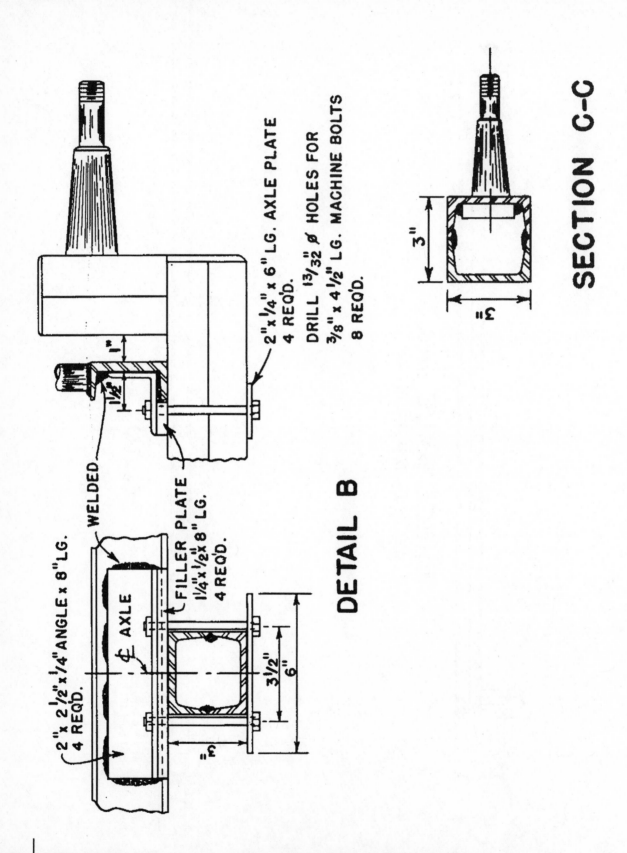

2" x 1/4" x 6" LG. AXLE PLATE
4 REQ'D.

DRILL 13/32" Ø HOLES FOR
3/8" x 4 1/2" LG. MACHINE BOLTS
8 REQ'D.

SECTION C-C

3"

3"

WELDED

℄ AXLE

FILLER PLATE
1/4" x 1/2" x 8" LG.
4 REQ'D.

2" x 2 1/2" x 1/4" ANGLE x 8" LG.
4 REQ'D.

3 1/2"

6"

3"

DETAIL B

1"

1 1/2"

The Complete Guide to Building Classic Barns, Fences, Storage Sheds,
Animal Pens, Outbuildings, Greenhouses, Farm Equipment, & Tools

• Seperator

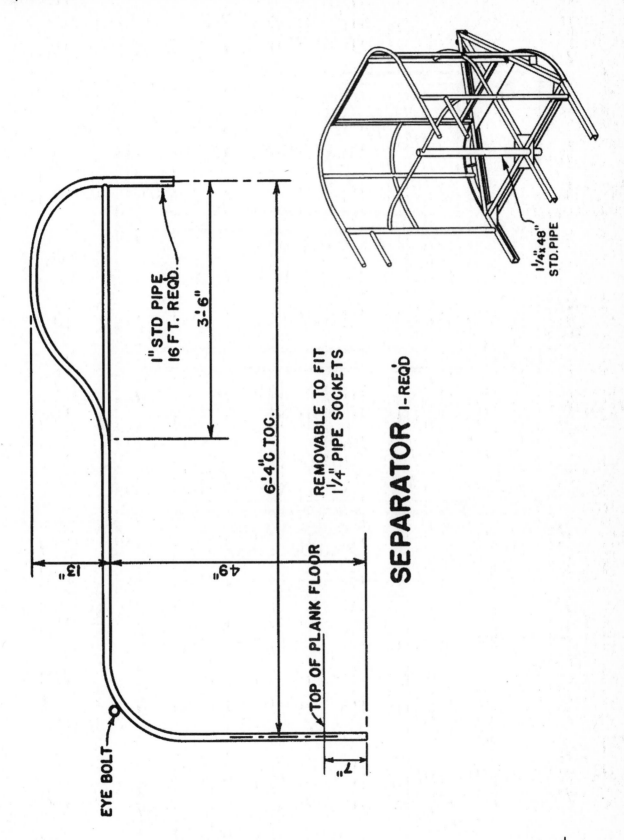

1" STD PIPE
16 FT. REQD.

3'-6"

6'-4" C TO C.

REMOVABLE TO FIT
1 1/4" PIPE SOCKETS

13"

49"

TOP OF PLANK FLOOR

EYE BOLT

7"

1 1/4" x 48"
STD. PIPE

SEPARATOR 1-REQD

• Tailgate

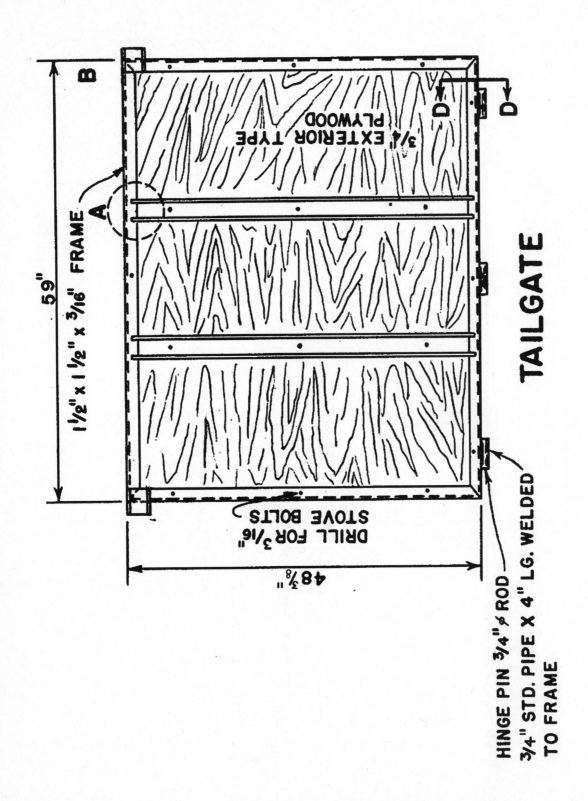

70

The Complete Guide to Building Classic Barns, Fences, Storage Sheds, Animal Pens, Outbuildings, Greenhouses, Farm Equipment, & Tools

• Tailgate Details A, B, & Section D-D

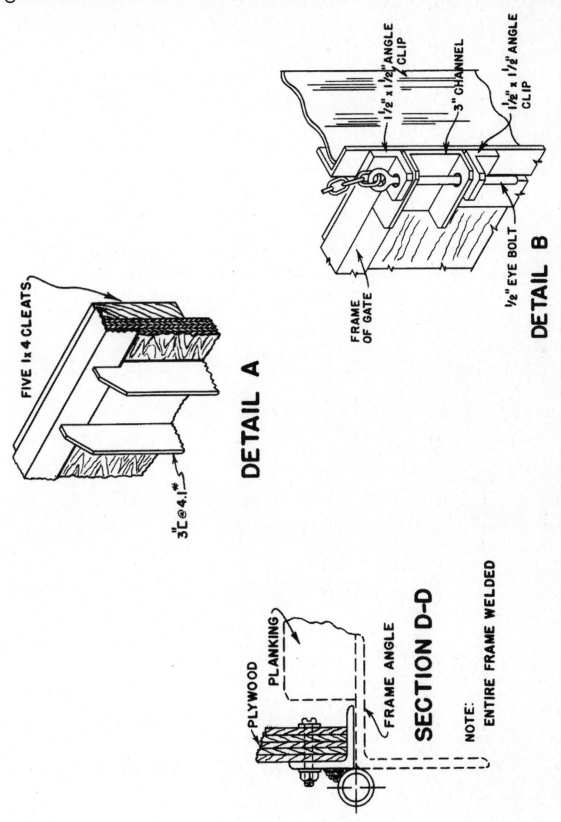

1½" x 1½" ANGLE CLIP

3" CHANNEL

1½" x 1½" ANGLE CLIP

½" EYE BOLT

FRAME OF GATE

DETAIL B

FIVE 1x4 CLEATS

3"C @ 4.1#

DETAIL A

PLYWOOD PLANKING

FRAME ANGLE

SECTION D-D

NOTE:

ENTIRE FRAME WELDED

THE COMPLETE GUIDE TO BUILDING CLASSIC BARNS, FENCES, STORAGE SHEDS, ANIMAL PENS, OUTBUILDINGS, GREENHOUSES, FARM EQUIPMENT, & TOOLS

chapter 4

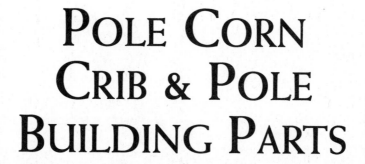

Pole Corn Crib & Pole Building Parts

• Plan • Perspective View

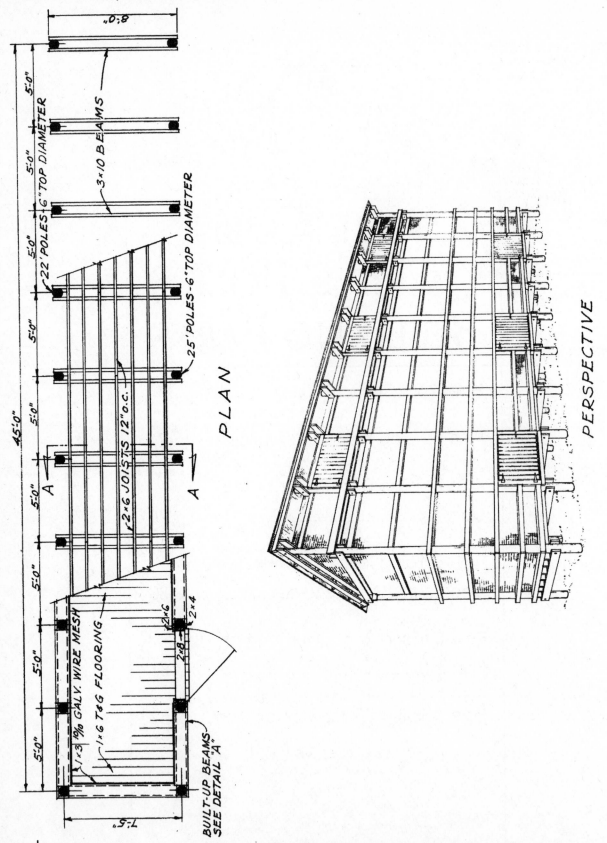

PLAN

PERSPECTIVE

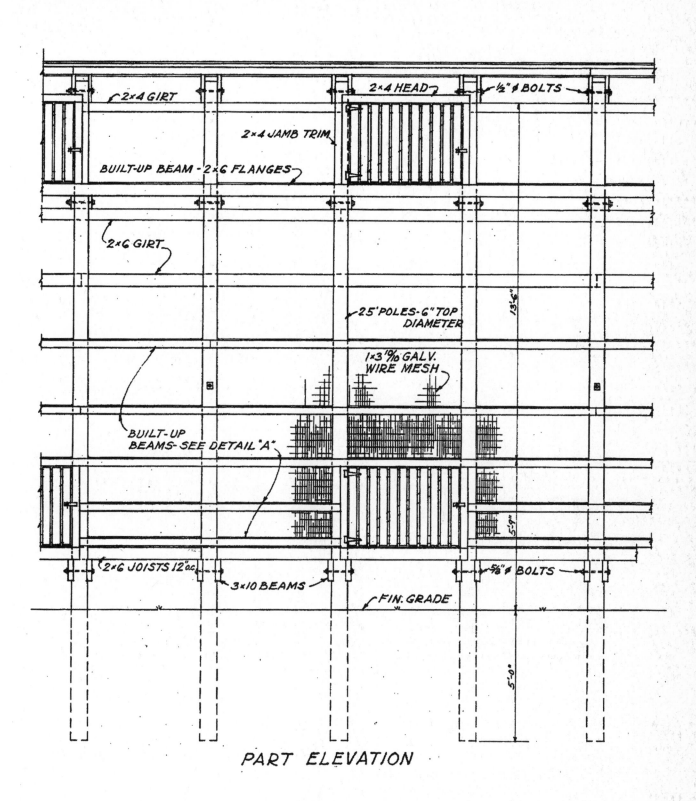

2×4 GIRT

2×4 HEAD

½"∅ BOLTS

2×4 JAMB TRIM

BUILT-UP BEAM - 2×6 FLANGES

2×6 GIRT

25' POLES - 6" TOP DIAMETER

13'-6"

1×3"⁹⁄₁₀ GALV. WIRE MESH

BUILT-UP BEAMS - SEE DETAIL "A"

5'-9"

2×6 JOISTS 12"oc

3×10 BEAMS

5⅝"∅ BOLTS

FIN. GRADE

5'-0"

PART ELEVATION

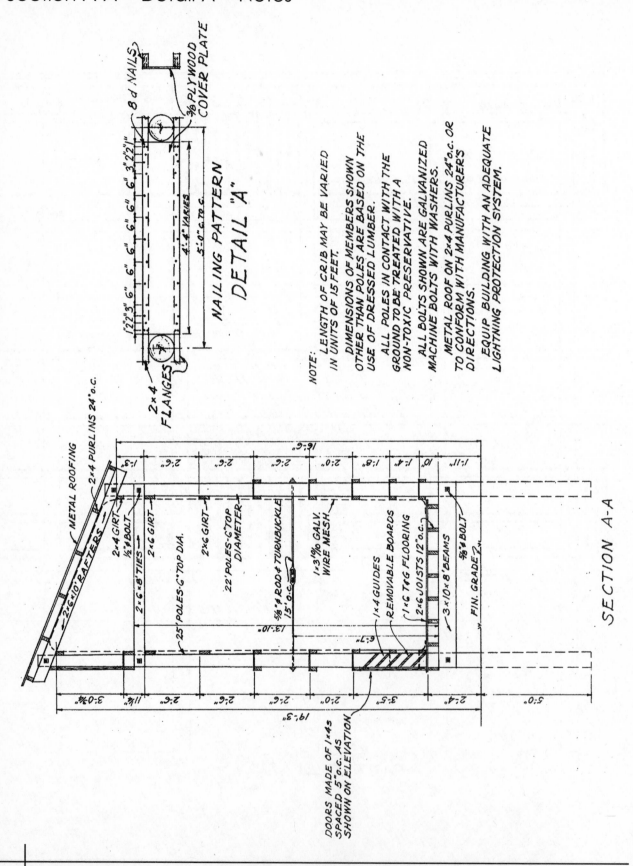

8 d NAILS

⅜ PLYWOOD COVER PLATE

NAILING PATTERN

DETAIL "A"

1" 2" 2" 3" 6" 6" 6" 6" 6" 6" 6" 3" 2" 2" 1"

4'-4" VARIES

5'-0" C. TO C.

2×4 FLANGES

NOTE: LENGTH OF CRIB MAY BE VARIED IN UNITS OF 15 FEET.

DIMENSIONS OF MEMBERS SHOWN OTHER THAN POLES ARE BASED ON THE USE OF DRESSED LUMBER.

ALL POLES IN CONTACT WITH THE GROUND TO BE TREATED WITH A NON-TOXIC PRESERVATIVE.

ALL BOLTS SHOWN ARE GALVANIZED MACHINE BOLTS WITH WASHERS.

METAL ROOF ON 2×4 PURLINS 24"o.c. OR TO CONFORM WITH MANUFACTURER'S DIRECTIONS.

EQUIP BUILDING WITH AN ADEQUATE LIGHTNING PROTECTION SYSTEM.

METAL ROOFING

2×4 PURLINS 24"o.c.

2×4 GIRT
½"⌀ BOLT

2×6 GIRT

2×6×8'TIES

2×6 GIRT

25'POLES-6"TOP DIA.

22'POLES-6"TOP DIAMETER

⅝"⌀ ROD & TURNBUCKLE 15'-0" o.c.

1×3½% GALV. WIRE MESH

1×4 GUIDES

REMOVABLE BOARDS

1×6 T&G FLOORING

2×6 JOISTS 12"o.c.

3×10×8 BEAMS

⅝"⌀ BOLT

FIN. GRADE

SECTION A-A

16'-6"

1'-3" 2'-6" 2'-6" 2'-6" 2'-0" 1'-8" 1'-4" 10" 1'-11"

13'-10"

9'-7"

3'-0¾" 11¼" 2'-6" 2'-6" 2'-6" 2'-0" 3'-5" 2'-4" 5'-0"

19'-3"

DOORS MADE OF 1×4s SPACED 5"o.c. AS SHOWN ON ELEVATION

2×6×10'RAFTERS

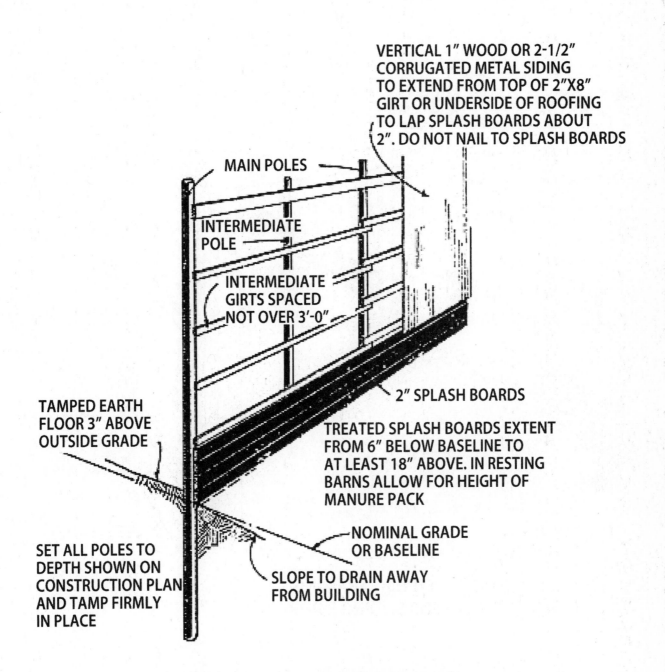

VERTICAL 1" WOOD OR 2-1/2"
CORRUGATED METAL SIDING
TO EXTEND FROM TOP OF 2"X8"
GIRT OR UNDERSIDE OF ROOFING
TO LAP SPLASH BOARDS ABOUT
2". DO NOT NAIL TO SPLASH BOARDS

MAIN POLES

INTERMEDIATE
POLE

INTERMEDIATE
GIRTS SPACED
NOT OVER 3'-0"

2" SPLASH BOARDS

TAMPED EARTH
FLOOR 3" ABOVE
OUTSIDE GRADE

TREATED SPLASH BOARDS EXTENT
FROM 6" BELOW BASELINE TO
AT LEAST 18" ABOVE. IN RESTING
BARNS ALLOW FOR HEIGHT OF
MANURE PACK

NOMINAL GRADE
OR BASELINE

SET ALL POLES TO
DEPTH SHOWN ON
CONSTRUCTION PLAN
AND TAMP FIRMLY
IN PLACE

SLOPE TO DRAIN AWAY
FROM BUILDING

TYPICAL WALL CONSTRUCTION
(ROOF FRAMING NOT SHOWN)

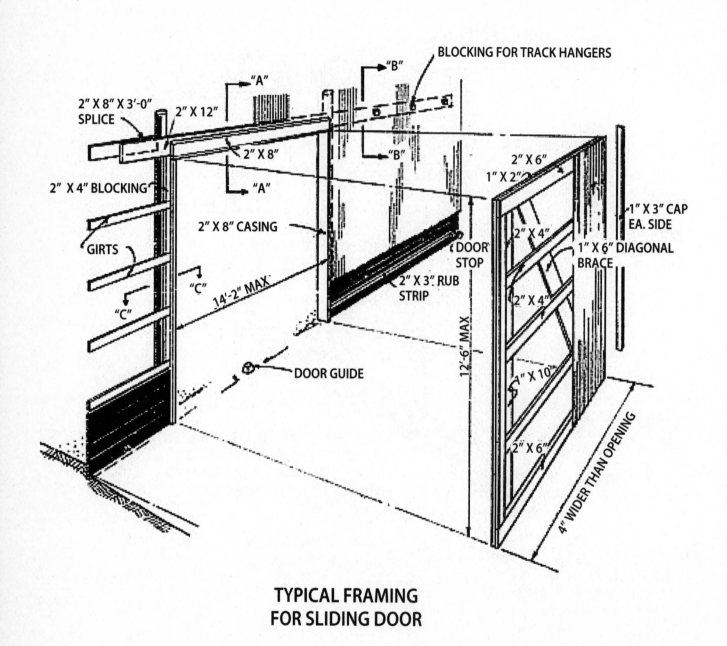

**TYPICAL FRAMING
FOR SLIDING DOOR**

• Sections A-A, B-B, & C-C

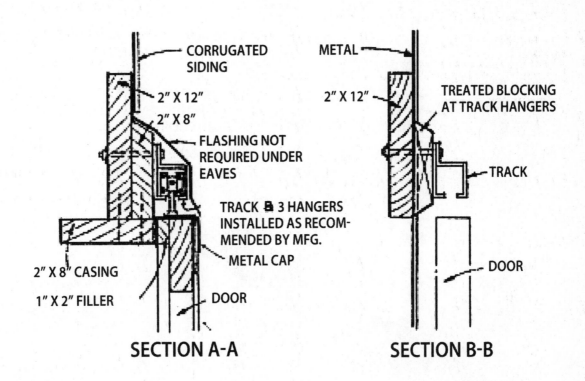

CORRUGATED
SIDING

2" X 12"

2" X 8"

FLASHING NOT
REQUIRED UNDER
EAVES

TRACK & 3 HANGERS
INSTALLED AS RECOM-
MENDED BY MFG.

METAL CAP

2" X 8" CASING

1" X 2" FILLER

DOOR

SECTION A-A

METAL

2" X 12"

TREATED BLOCKING
AT TRACK HANGERS

TRACK

DOOR

SECTION B-B

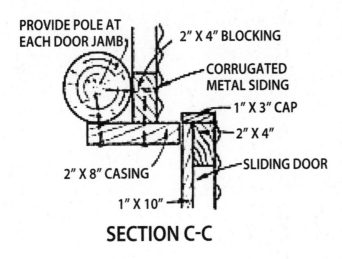

PROVIDE POLE AT
EACH DOOR JAMB

2" X 4" BLOCKING

CORRUGATED
METAL SIDING

1" X 3" CAP

2" X 4"

SLIDING DOOR

2" X 8" CASING

1" X 10"

SECTION C-C

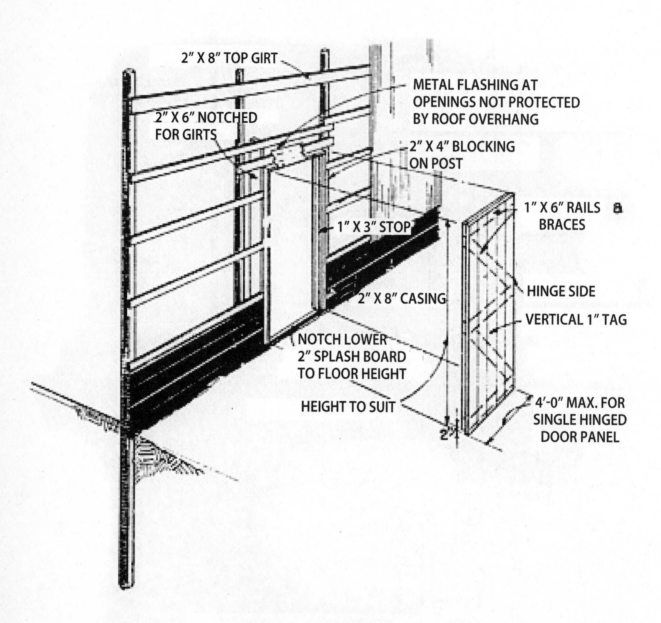

2" X 8" TOP GIRT

METAL FLASHING AT OPENINGS NOT PROTECTED BY ROOF OVERHANG

2" X 6" NOTCHED FOR GIRTS

2" X 4" BLOCKING ON POST

1" X 6" RAILS & BRACES

1" X 3" STOP

HINGE SIDE

VERTICAL 1" TAG

2" X 8" CASING

NOTCH LOWER 2" SPLASH BOARD TO FLOOR HEIGHT

HEIGHT TO SUIT

2"

4'-0" MAX. FOR SINGLE HINGED DOOR PANEL

TYPICAL FRAMING FOR HINGED DOOR

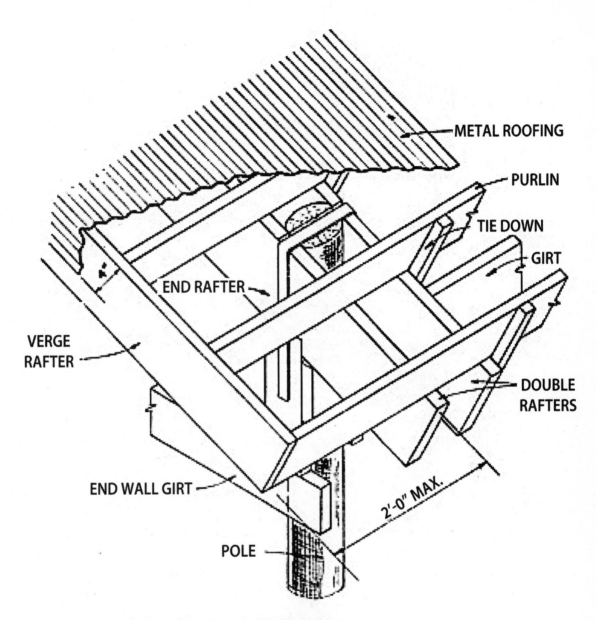

ROOF OVERHANG AT ENDWALL
(DOUBLE RAFTERS & PURLIN CONSTRUCTION SHOWN)
WHEN SHEATHING IS USED INSTEAD OF PURLINS IT MAY BE
EXTENDED IN SIMILIAR MANNER, PROVIDE BLOCKING BETWEEN
END & VERGE RAFTERS IF OVERHANG EXCEEDS 12".

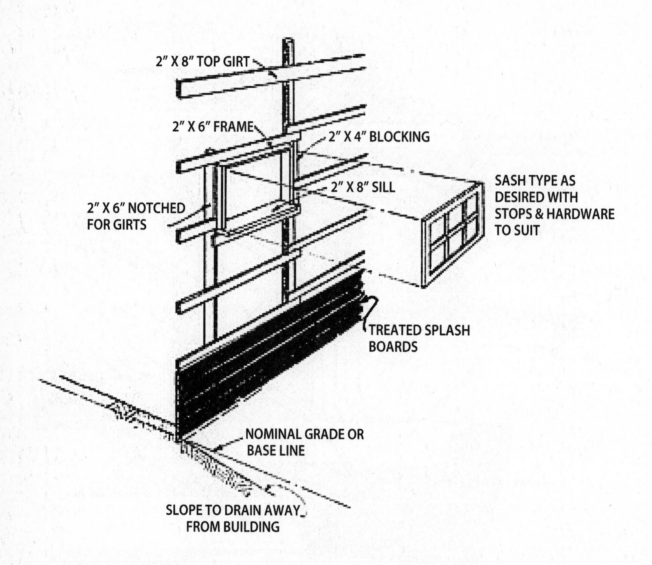

2" X 8" TOP GIRT

2" X 6" FRAME

2" X 4" BLOCKING

2" X 8" SILL

SASH TYPE AS DESIRED WITH STOPS & HARDWARE TO SUIT

2" X 6" NOTCHED FOR GIRTS

TREATED SPLASH BOARDS

NOMINAL GRADE OR BASE LINE

SLOPE TO DRAIN AWAY FROM BUILDING

TYPICAL WINDOW INSTALLATION

chapter 5

STORAGE SHEDS

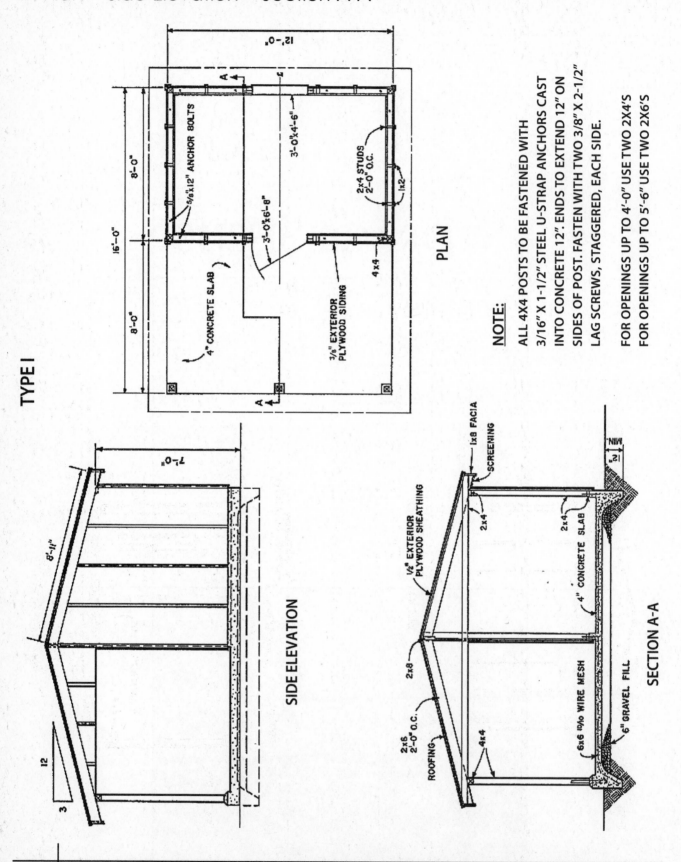

TYPE I

PLAN

12'-0"

16'-0"

8'-0"

8'-0"

5/8"x12" ANCHOR BOLTS

3'-0"x4'-6"

2x4 STUDS 2'-0" O.C.

1x2

3'-0"x6'-8"

4x4

4" CONCRETE SLAB

3/8" EXTERIOR PLYWOOD SIDING

NOTE:

ALL 4X4 POSTS TO BE FASTENED WITH 3/16" X 1-1/2" STEEL U-STRAP ANCHORS CAST INTO CONCRETE 12". ENDS TO EXTEND 12" ON SIDES OF POST. FASTEN WITH TWO 3/8" X 2-1/2" LAG SCREWS, STAGGERED, EACH SIDE.

FOR OPENINGS UP TO 4'-0" USE TWO 2X4'S
FOR OPENINGS UP TO 5'-6" USE TWO 2X6'S

SIDE ELEVATION

7'-0"

8'-11"

12
3

SECTION A-A

1x8 FACIA

SCREENING

1/2" EXTERIOR PLYWOOD SHEATHING

2x4

2x4

4" CONCRETE SLAB

12" MIN

2x8

6x6 10/10 WIRE MESH

6" GRAVEL FILL

4x4

2x6 2'-0" O.C.

ROOFING

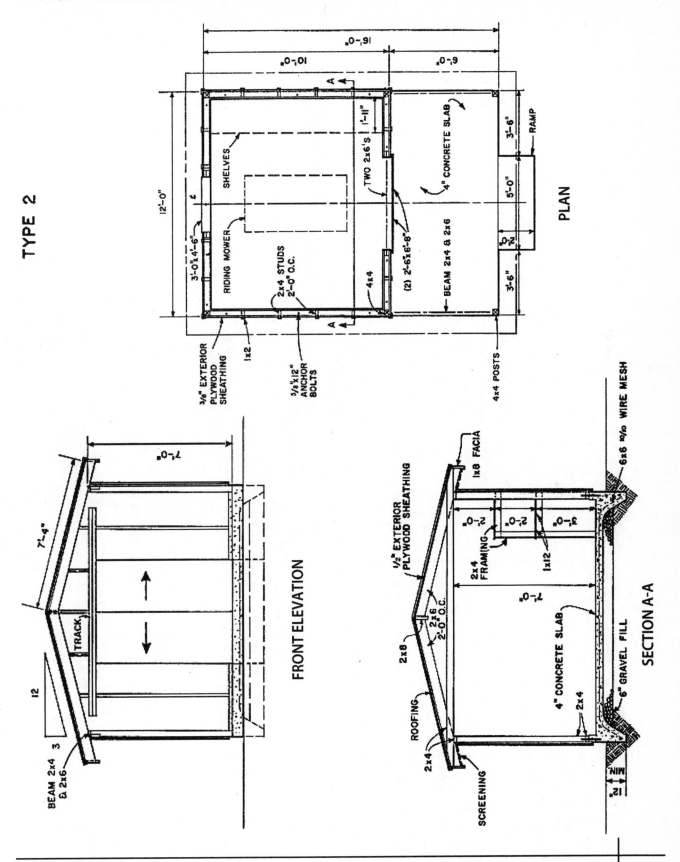

TYPE 2

PLAN

FRONT ELEVATION

SECTION A-A

STORAGE SHED III

• Plan • Front Elevation • Section A-A

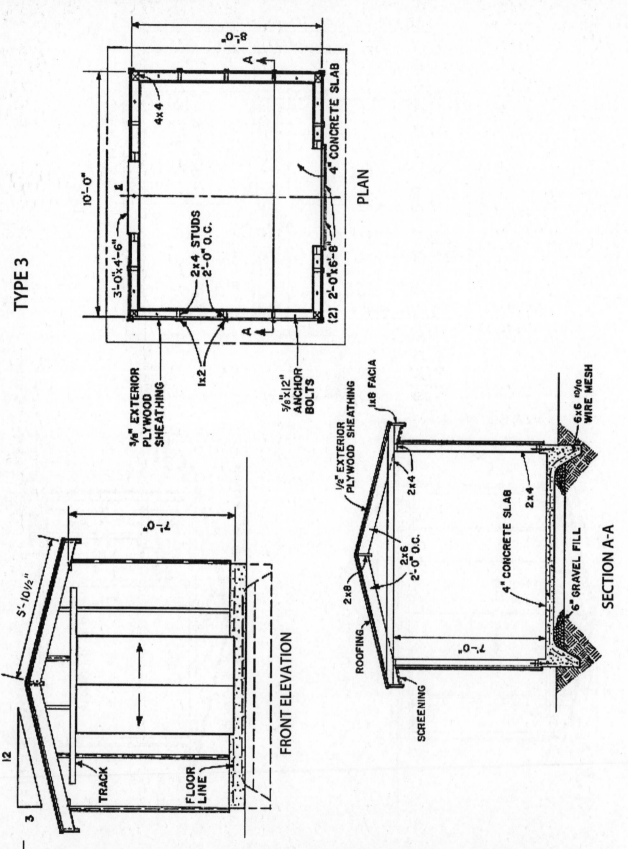

TYPE 3

8'-0"

10'-0"

4" CONCRETE SLAB

PLAN

4x4

3'-0" x 4'-6"

2x4 STUDS 2'-0" O.C.

1x2

(2) 2'-0" x 6'-8"

3/8" EXTERIOR PLYWOOD SHEATHING

5/8" x 12" ANCHOR BOLTS

7'-0"

5'-10½"

12

3

TRACK

FLOOR LINE

FRONT ELEVATION

1x8 FACIA

½" EXTERIOR PLYWOOD SHEATHING

2x4

2x6 2'-0" O.C.

2x8

ROOFING

SCREENING

4" CONCRETE SLAB

2x4

6x6 10/10 WIRE MESH

6" GRAVEL FILL

7'-0"

SECTION A-A

THE COMPLETE GUIDE TO BUILDING CLASSIC BARNS, FENCES, STORAGE SHEDS, ANIMAL PENS, OUTBUILDINGS, GREENHOUSES, FARM EQUIPMENT, & TOOLS

CASE STUDY: STORAGE SHEDS I

Gen Sterenberg
Grand Rapids, Michigan

Gen Sterenberg and her husband built a 12-foot by 16-foot storage shed in her Michigan backyard that she uses to store items that just won't fit in her garage. Sterenburg said she and her husband decided on that size for their shed because they wanted to keep the building small enough that it did not need a building permit. She and her husband originally planned to do the work themselves, but when their friend who is a licensed builder offered to build it for them, they took him up on his offer and worked alongside him.

"We were very happy with our building," she said. "We wish we had made it larger, but then is it ever big enough? We really didn't want to take up more yard space, and we didn't want to deal with building permits."

Sterenberg's shed houses garden hoses, lawn sprinklers, shovels, rakes, a rototiller, boating equipment, large carpentry tools, beach toys, and other things that just don't seem to have a proper place of their own. Although the shed currently only has four walls and a concrete slab foundation, she hopes to install freestanding shelves to offer more storage options. The shed isn't wired for electricity, which is acceptable most of the time. In those instances where the Sterenbergs need a light source, they run an extension cord from the house.

The advice she offers to other homeowners looking to build a shed or other building on their property is to check the zoning and building codes for their area. She also suggests that homeowners opt to build a larger structure as opposed to a small one.

"Build large because whatever size it is, it's not big enough!"

CLASSIFIED CASE STUDIES
directly from the experts

In 2009, Thomas Kocourek built a 25-foot by 25-foot two-car garage and workshop, which he attached to a heated workshop on his 5-acre hobby farm. Because of the work he does on his hobby farm, Kocourek said he is always repairing some piece of equipment or working on a small project and said he'd need a larger heated workspace. He hired a contractor to work on the structure, and Kocourek did the electrical work on the building himself. He said he is pleased with the results of his building, but wishes the space was larger. He is planning to build a 40-foot by 60-foot outbuilding in the next few years to store several big boats.

"I think the biggest mistake people make when building is they always wish it was bigger," Kocourek said. "Often, limited funds are the reason for not building bigger or not using the most energy efficient designs."

The first piece of advice Kocourek offers is to build large because he said homeowners will always wish they had more storage space. He also recommends that homeowners try to incorporate as many energy efficient designs as possible because energy costs will only continue to climb in the future.

In addition to the structure he built, Kocourek has a 105-year old barn on his property. He said he is always amazed at the engineering that went into building the structure and how well it has stood the test of time given the limited tools available when the building was constructed. He put on a new roof only five years ago and is replacing the weathered boards that have lasted more than 100 years.

"Old barns are a great piece of history. It is often cheaper in the long run to tear down those old barns and build new [structures], but preserving these great old historical buildings is worth the effort and added cost," he said.

chapter 6

SMALL FARM ANIMAL BUILDINGS

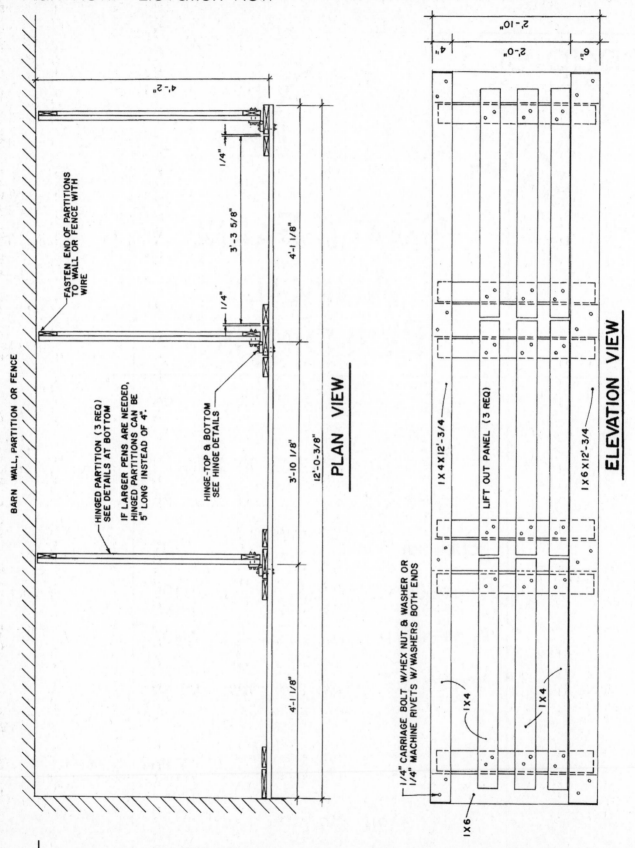

PLAN VIEW

ELEVATION VIEW

BARN WALL, PARTITION OR FENCE

FASTEN END OF PARTITIONS TO WALL OR FENCE WITH WIRE

HINGED PARTITION (3 REQ) SEE DETAILS AT BOTTOM

IF LARGER PENS ARE NEEDED, HINGED PARTITIONS CAN BE 5" LONG INSTEAD OF 4".

HINGE-TOP & BOTTOM SEE HINGE DETAILS

4'-2"

3'-3 5/8"

4'-1 1/8"

1/4"

1/4"

3'-10 1/8"

12'-0-3/8"

4'-1 1/8"

2'-10"

4"

2'-0"

6"

1 X 4 X 12'-3/4

LIFT OUT PANEL (3 REQ)

1 X 6 X 12'-3/4

1/4" CARRIAGE BOLT W/HEX NUT & WASHER OR 1/4" MACHINE RIVETS W/WASHERS BOTH ENDS

1 X 4

1 X 4

1 X 6

• Lift Out Panel Details • Partition Details

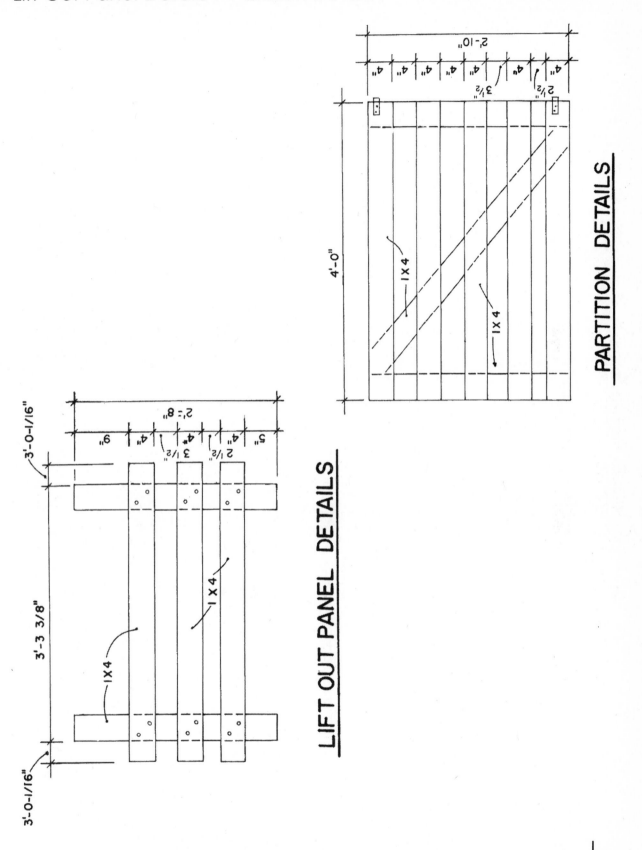

PARTITION DETAILS

LIFT OUT PANEL DETAILS

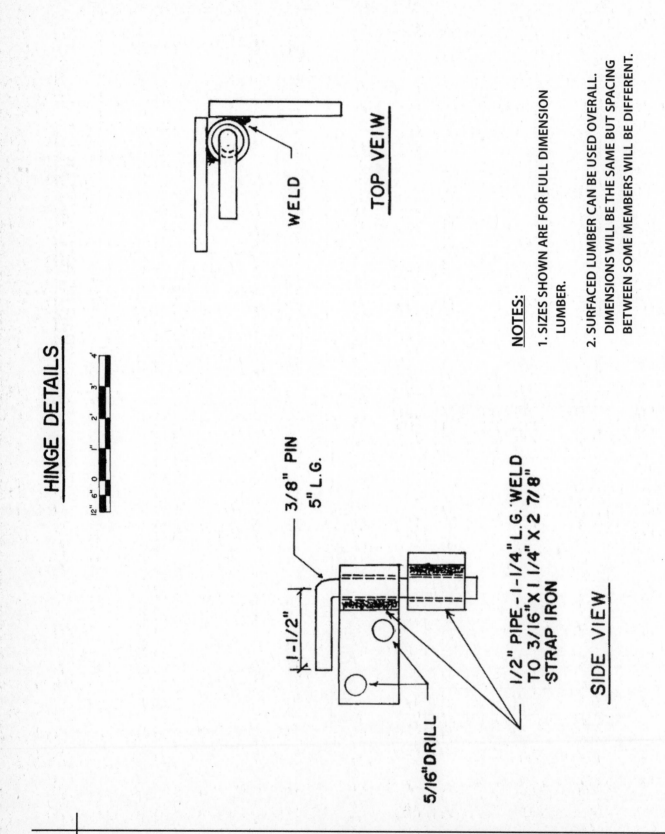

HINGE DETAILS

12" 6" 0 1' 2' 3' 4'

WELD

TOP VEIW

3/8" PIN
5" L.G.

1-1/2"

5/16" DRILL

1/2" PIPE – 1-1/4" L.G. WELD
TO 3/16" X 1 1/4" X 2 7/8"
STRAP IRON

SIDE VIEW

<u>NOTES:</u>

1. SIZES SHOWN ARE FOR FULL DIMENSION LUMBER.

2. SURFACED LUMBER CAN BE USED OVERALL. DIMENSIONS WILL BE THE SAME BUT SPACING BETWEEN SOME MEMBERS WILL BE DIFFERENT.

10X12 POULTRY HOUSE

• Plan • Perspective View

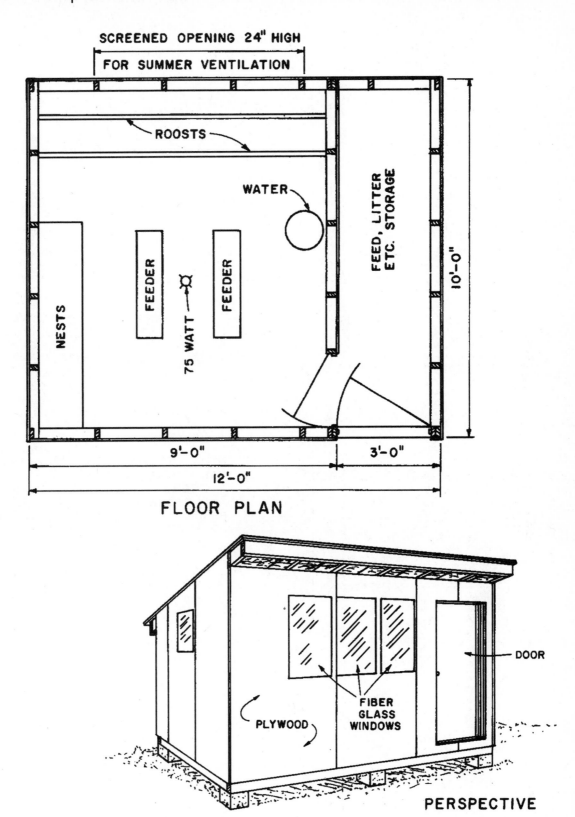

SCREENED OPENING 24" HIGH
FOR SUMMER VENTILATION

ROOSTS

WATER

FEED, LITTER ETC. STORAGE

NESTS

FEEDER

75 WATT

FEEDER

10'-0"

9'-0"

3'-0"

12'-0"

FLOOR PLAN

PLYWOOD

FIBER GLASS WINDOWS

DOOR

PERSPECTIVE

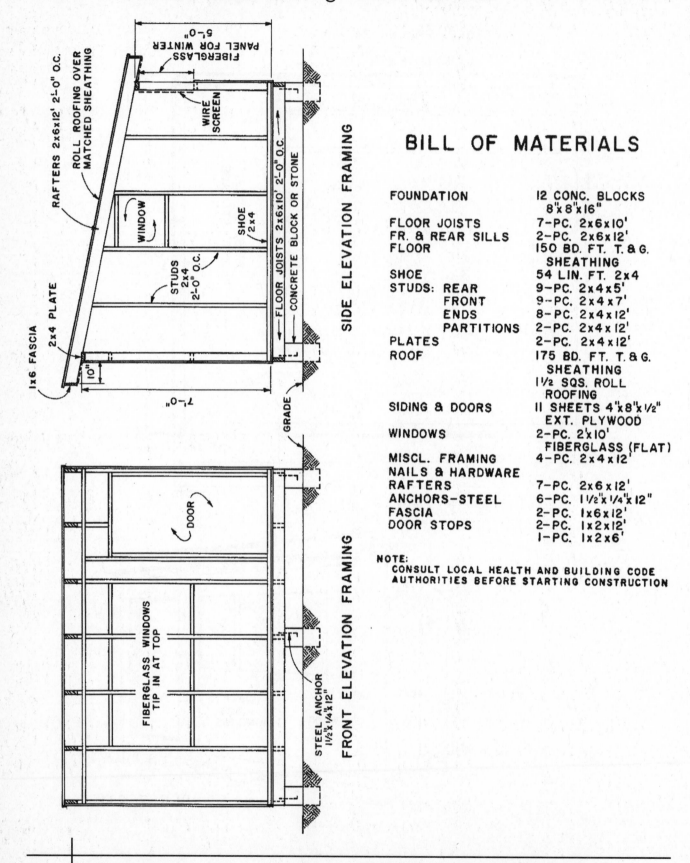

RAFTERS 2x6x12' 2'-0" O.C.

ROLL ROOFING OVER MATCHED SHEATHING

FIBERGLASS PANEL FOR WINTER 5'-0"

WIRE SCREEN

WINDOW

SHOE 2x4

STUDS 2x4 2'-0" O.C.

FLOOR JOISTS 2x6x10' 2'-0" O.C.

CONCRETE BLOCK OR STONE

2x4 PLATE

1x6 FASCIA

10"

7'-0"

GRADE

SIDE ELEVATION FRAMING

FIBERGLASS WINDOWS TIP IN AT TOP

DOOR

STEEL ANCHOR 1½"x ¼"x 12"

FRONT ELEVATION FRAMING

BILL OF MATERIALS

FOUNDATION	12 CONC. BLOCKS 8"x 8"x16"
FLOOR JOISTS	7-PC. 2x6x10'
FR. & REAR SILLS	2-PC. 2x6x12'
FLOOR	150 BD. FT. T. & G. SHEATHING
SHOE	54 LIN. FT. 2x4
STUDS: REAR	9-PC. 2x4x5'
FRONT	9-PC. 2x4x7'
ENDS	8-PC. 2x4x12'
PARTITIONS	2-PC. 2x4x12'
PLATES	2-PC. 2x4x12'
ROOF	175 BD. FT. T. & G. SHEATHING 1½ SQS. ROLL ROOFING
SIDING & DOORS	11 SHEETS 4"x8"x½" EXT. PLYWOOD
WINDOWS	2-PC. 2'x10' FIBERGLASS (FLAT)
MISCL. FRAMING	4-PC. 2x4x12'
NAILS & HARDWARE	
RAFTERS	7-PC. 2x6x12'
ANCHORS-STEEL	6-PC. 1½"x¼"x12"
FASCIA	2-PC. 1x6x12'
DOOR STOPS	2-PC. 1x2x12' 1-PC. 1x2x6'

NOTE:
CONSULT LOCAL HEALTH AND BUILDING CODE AUTHORITIES BEFORE STARTING CONSTRUCTION

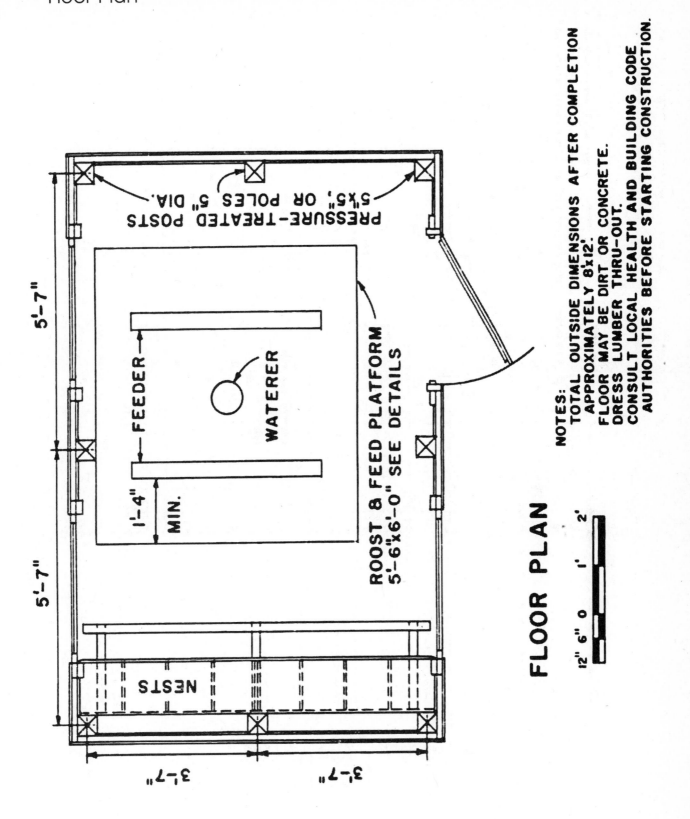

PRESSURE-TREATED POSTS
5"x5" OR POLES 5" DIA.

5'-7"

5'-7"

FEEDER

WATERER

1'-4"

MIN.

ROOST & FEED PLATFORM
5'-6"x6'-0" SEE DETAILS

NESTS

3'-7"

3'-7"

FLOOR PLAN

12" 6" 0   1'   2'

NOTES:
TOTAL OUTSIDE DIMENSIONS AFTER COMPLETION
APPROXIMATELY 8'x12'.
FLOOR MAY BE DIRT OR CONCRETE.
DRESS LUMBER THRU-OUT.
CONSULT LOCAL HEALTH AND BUILDING CODE
AUTHORITIES BEFORE STARTING CONSTRUCTION.

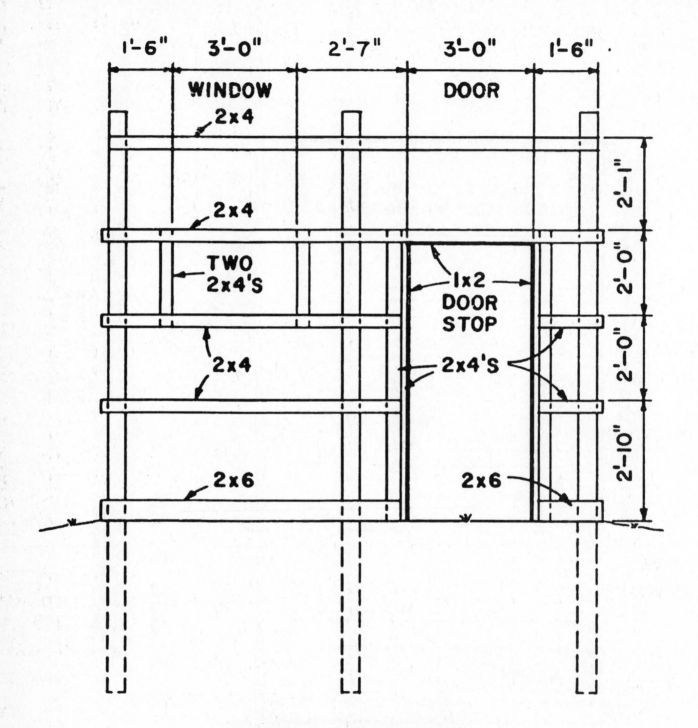

FRONT FRAMING

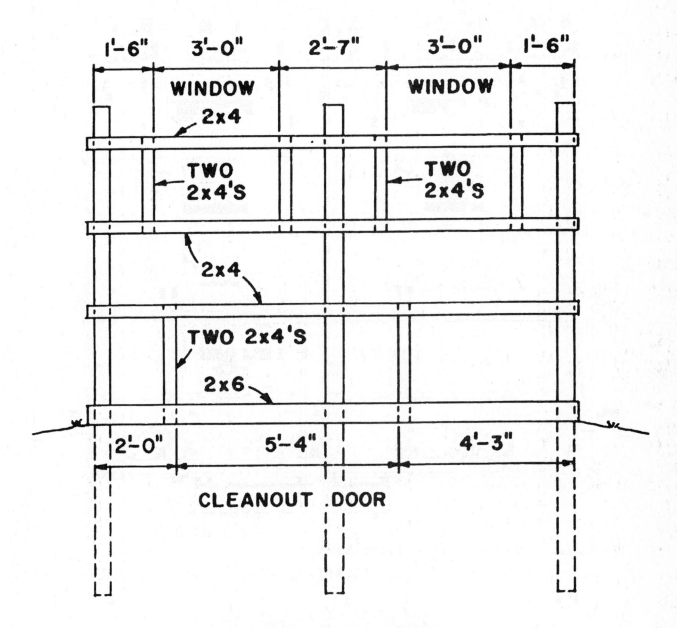

1'-6" 3'-0" 2'-7" 3'-0" 1'-6"

WINDOW
2x4

WINDOW

TWO
2x4'S

TWO
2x4'S

2x4

TWO 2x4'S

2x6

2'-0" 5'-4" 4'-3"

CLEANOUT DOOR

REAR FRAMING

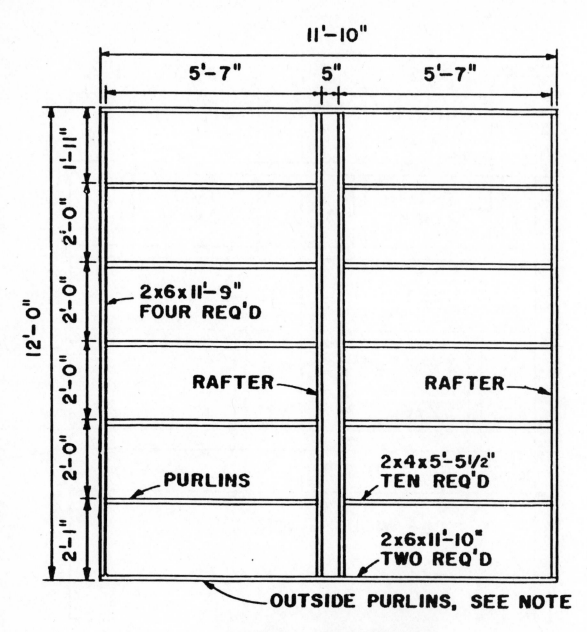

ROOF FRAMING

NOTE:
BUILD ROOF FRAME IN TWO PANELS.
NAIL EACH PANEL TO P.T. POSTS, THEN
ATTACH THE TWO OUTSIDE PURLINS.

• Side Framing

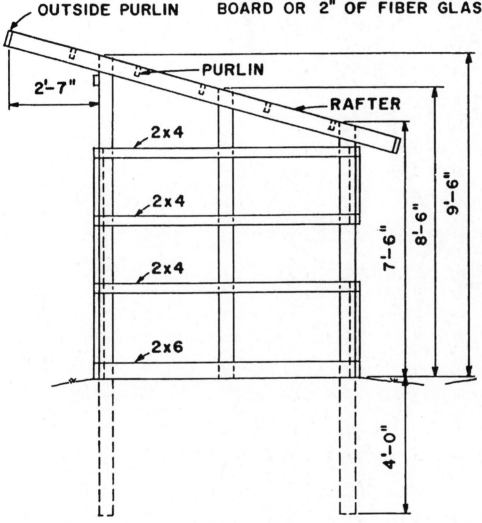

ROOF MAY BE METAL OR ½" EXT. PLYWOOD OR OTHER SHEATHING WITH SHINGLES OR ROLL ROOFING. INSULATE ROOF WITH INSULATION BOARD OR 2" OF FIBER GLASS.

OUTSIDE PURLIN

2'-7"

PURLIN

RAFTER

2x4

2x4

2x4

2x6

9'-6"

8'-6"

7'-6"

4'-0"

SIDE FRAMING

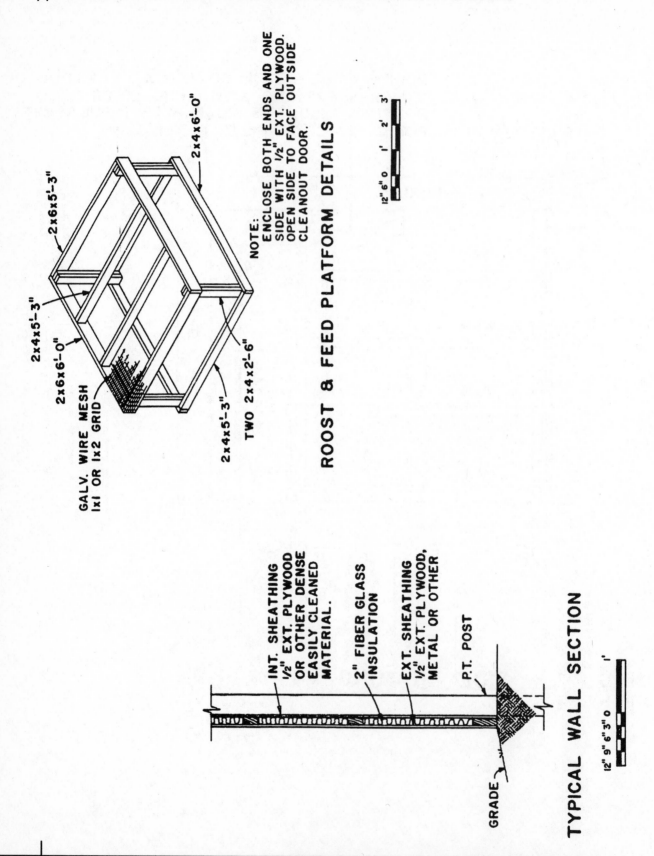

2x6x5'-3"

2x4x6'-0"

NOTE:
ENCLOSE BOTH ENDS AND ONE
SIDE WITH ½" EXT. PLYWOOD.
OPEN SIDE TO FACE OUTSIDE
CLEANOUT DOOR.

2x4x5'-3"

2x6x6'-0"

GALV. WIRE MESH
1x1 OR 1x2 GRID

2x4x5'-3"

TWO 2x4x2'-6"

ROOST & FEED PLATFORM DETAILS

12" 6" 0   1'   2'   3'

INT. SHEATHING
½" EXT. PLYWOOD
OR OTHER DENSE
EASILY CLEANED
MATERIAL.

2" FIBER GLASS
INSULATION

EXT. SHEATHING
½" EXT. PLYWOOD,
METAL OR OTHER

P.T. POST

GRADE

TYPICAL WALL SECTION

12" 9" 6" 3" 0   1'

100

THE COMPLETE GUIDE TO BUILDING CLASSIC BARNS, FENCES, STORAGE SHEDS,
ANIMAL PENS, OUTBUILDINGS, GREENHOUSES, FARM EQUIPMENT, & TOOLS

• Exploded View

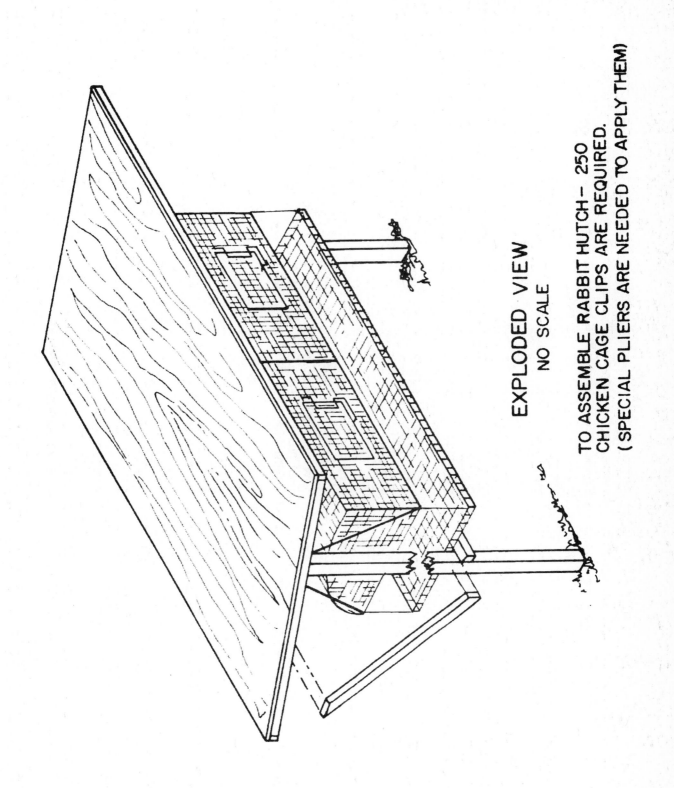

EXPLODED VIEW
NO SCALE

TO ASSEMBLE RABBIT HUTCH– 250
CHICKEN CAGE CLIPS ARE REQUIRED.
(SPECIAL PLIERS ARE NEEDED TO APPLY THEM)

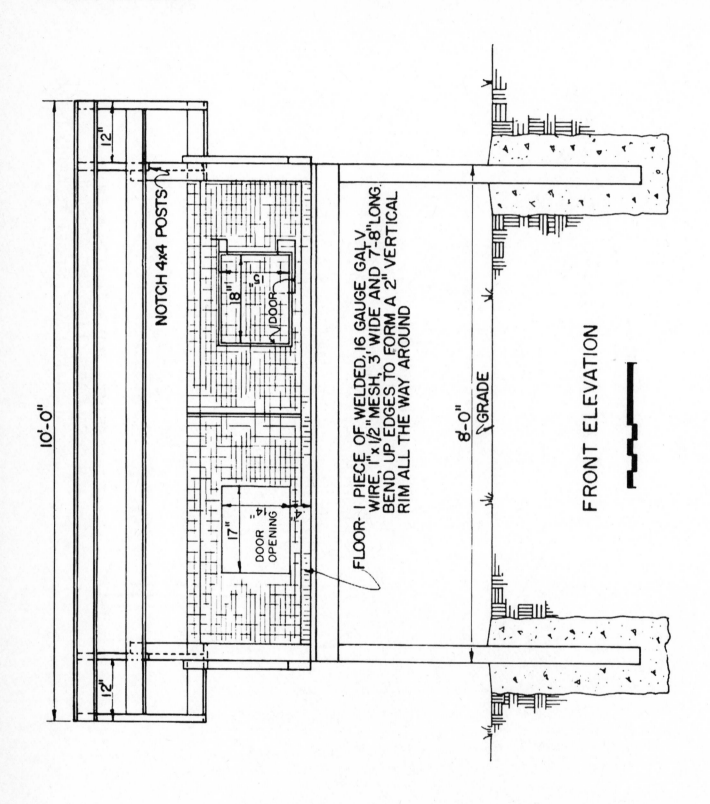

10'-0"

12"

12"

NOTCH 4x4 POSTS

18"

15"

DOOR

17"

14"

4"

DOOR
OPENING

FLOOR- 1 PIECE OF WELDED, 16 GAUGE GALV.
WIRE, 1"x 1/2" MESH, 3' WIDE AND 7-8" LONG.
BEND UP EDGES TO FORM A 2" VERTICAL
RIM ALL THE WAY AROUND

8'-0"

GRADE

FRONT ELEVATION

Rabbit Hutch

• Side Elevation • Latch Detail • Pattern for Side of Cage and Center Partition

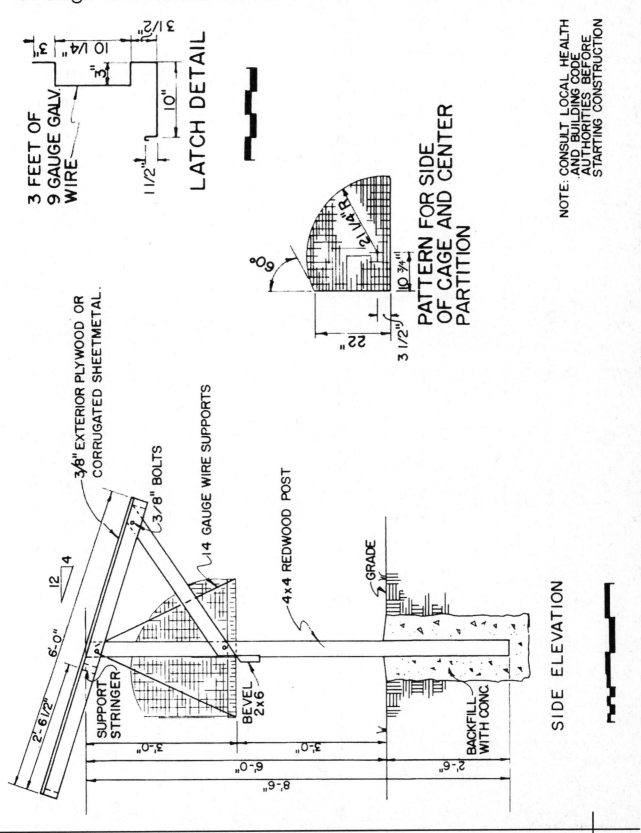

LATCH DETAIL

3 FEET OF 9 GAUGE GALV. WIRE

PATTERN FOR SIDE OF CAGE AND CENTER PARTITION

NOTE: CONSULT LOCAL HEALTH AND BUILDING CODE AUTHORITIES BEFORE STARTING CONSTRUCTION

SIDE ELEVATION

3/8" EXTERIOR PLYWOOD OR CORRUGATED SHEETMETAL.

3/8" BOLTS

14 GAUGE WIRE SUPPORTS

4x4 REDWOOD POST

GRADE

SUPPORT STRINGER

BEVEL 2x6

BACKFILL WITH CONC.

RABBIT HUTCH

• Wire Mesh Cutting Diagram

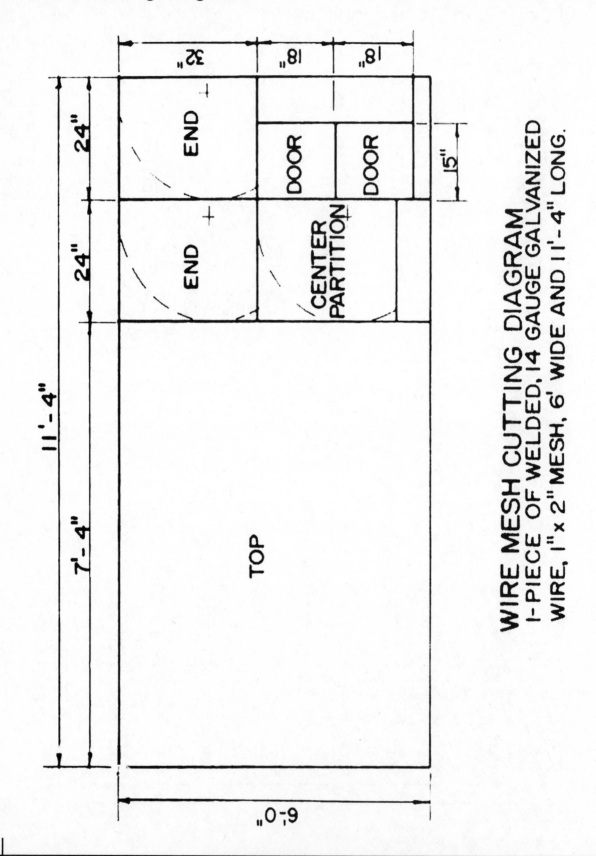

WIRE MESH CUTTING DIAGRAM
1-PIECE OF WELDED, 14 GAUGE GALVANIZED
WIRE, 1" x 2" MESH, 6' WIDE AND 11'- 4" LONG.

THE COMPLETE GUIDE TO BUILDING CLASSIC BARNS, FENCES, STORAGE SHEDS,
ANIMAL PENS, OUTBUILDINGS, GREENHOUSES, FARM EQUIPMENT, & TOOLS

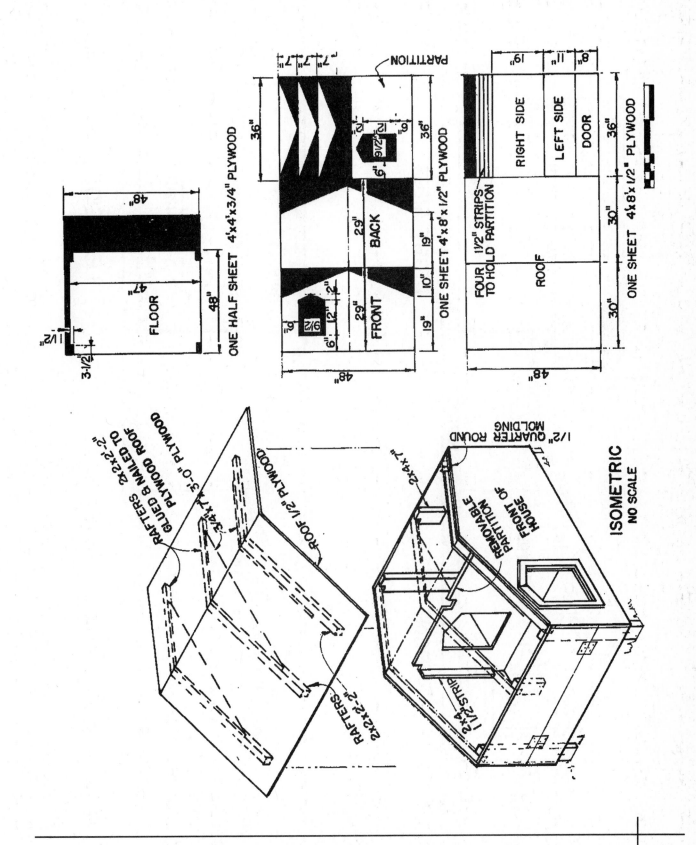

FLOOR

48"

47"

48"

1-1/2"

3-1/2"

ONE HALF SHEET 4'x4'x3/4" PLYWOOD

PARTITION

7" 7" 7" 7"

36"

36"

6" 9-1/2"

BACK

29"

19"

FRONT

29"

12" 2"

6" 9-1/2"

10"

19"

48"

ONE SHEET 4'x8'x1/2" PLYWOOD

9"

11"

8"

RIGHT SIDE

LEFT SIDE

DOOR

FOUR 1-1/2" STRIPS TO HOLD PARTITION

ROOF

36"

30"

30"

48"

ONE SHEET 4'x8'x1/2" PLYWOOD

RAFTERS NAILED TO GLUED PLYWOOD ROOF 2x2x2'-2"

ROOF 1/2" PLYWOOD

3/4"x1-1/4"

3'-0" PLYWOOD

RAFTERS 2x2x2'-2"

1/2" QUARTER ROUND MOLDING

2x4x7"

REMOVABLE PARTITION OF FRONT OF HOUSE

2x4"STRIP 1-1/2"

ISOMETRIC
NO SCALE

DOG HOUSE I

• Top, Front, Left, & Right Views

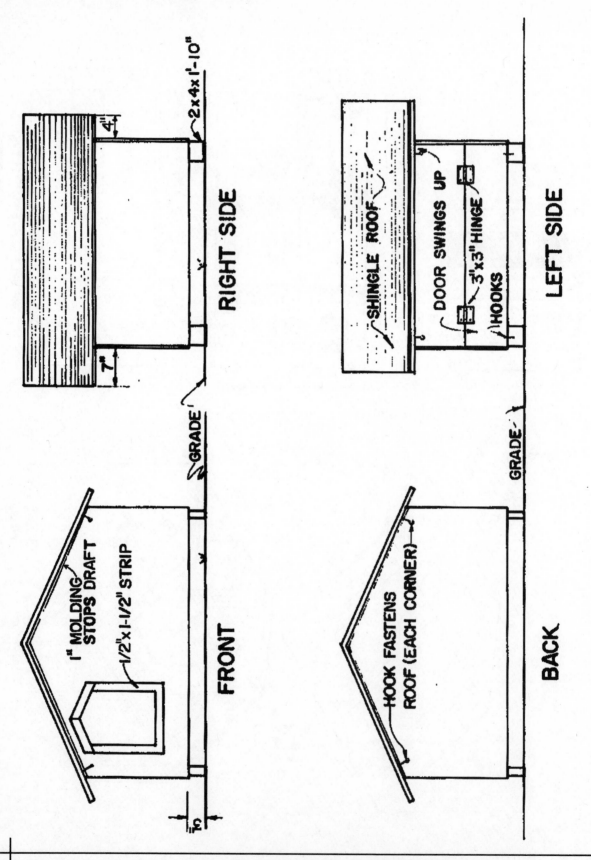

2x4x1'-10"

4"

RIGHT SIDE

7"

GRADE

SHINGLE ROOF

DOOR SWINGS UP

3"x3" HINGE

HOOKS

LEFT SIDE

GRADE

1" MOLDING STOPS DRAFT

1/2"x1-1/2" STRIP

FRONT

HOOK FASTENS ROOF (EACH CORNER)

BACK

THE COMPLETE GUIDE TO BUILDING CLASSIC BARNS, FENCES, STORAGE SHEDS, ANIMAL PENS, OUTBUILDINGS, GREENHOUSES, FARM EQUIPMENT, & TOOLS

DOG HOUSE II

• Top, Front, & Side Views • Plywood Cutouts

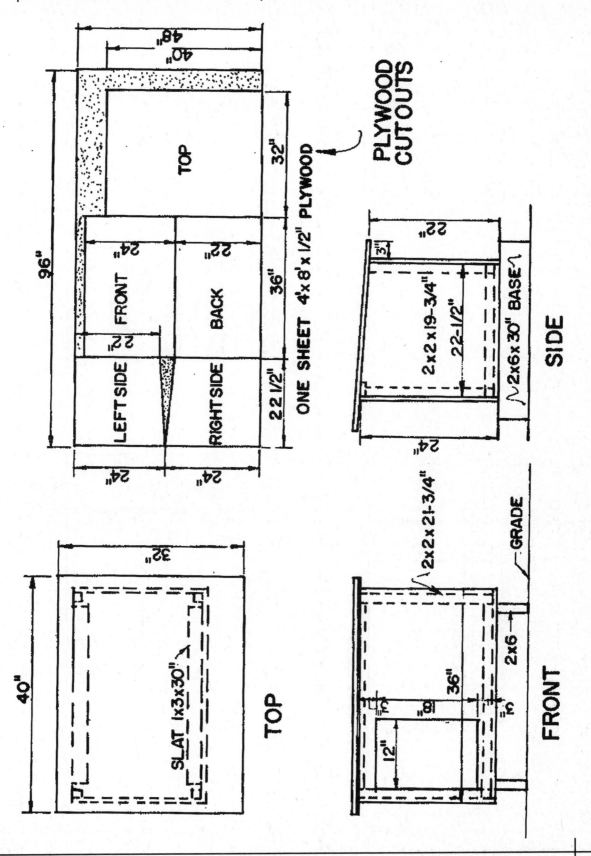

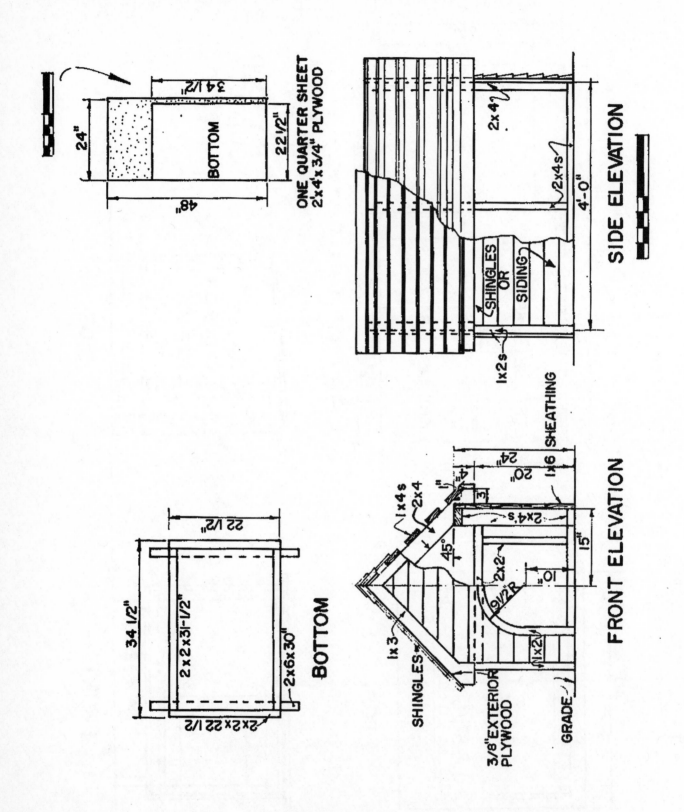

ONE QUARTER SHEET
2'x4'x 3/4" PLYWOOD

34 1/2"

24"

22 1/2"

48"

BOTTOM

SIDE ELEVATION

2x4's

2x4s

4'-0"

SHINGLES OR SIDING

1x2s

FRONT ELEVATION

1x6 SHEATHING

1x4s

2x4

4"

3"

24"

20"

2x4's

45°

2x2

9 1/2"R

15"

10"

1x3

SHINGLES

3/8" EXTERIOR PLYWOOD

1x2's

GRADE

BOTTOM

22 1/2"

34 1/2"

2x2x31-1/2"

2x6x30"

2x2x22 1/2

chapter 7

FEEDERS

Two Wheel Self-Feeder for Cattle
• Perspective View • Top View • Side View • Section A-A

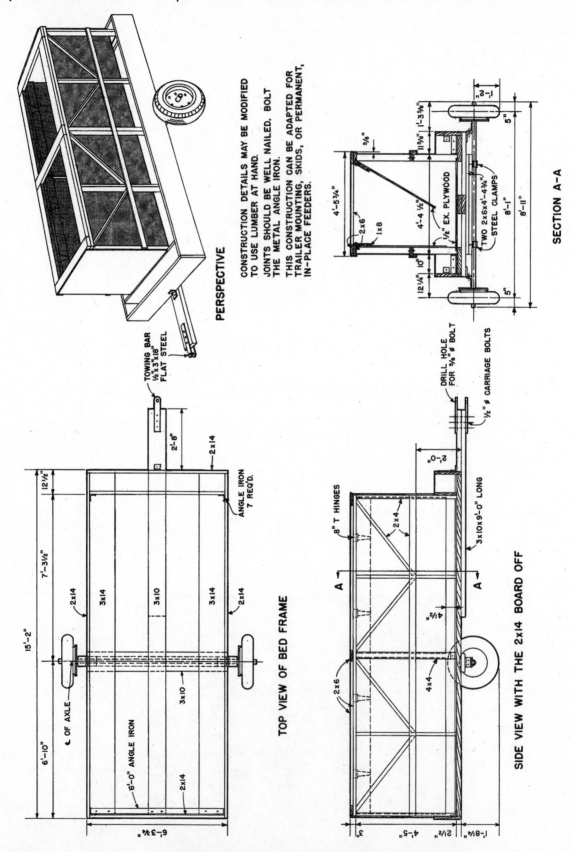

PERSPECTIVE

CONSTRUCTION DETAILS MAY BE MODIFIED TO USE LUMBER AT HAND. JOINTS SHOULD BE WELL NAILED. BOLT THE METAL ANGLE IRON.

THIS CONSTRUCTION CAN BE ADAPTED FOR TRAILER MOUNTING, SKIDS, OR PERMANENT, IN-PLACE FEEDERS.

SECTION A-A

1/2" EX. PLYWOOD

TWO 2x6x4-4 3/4" STEEL CLAMPS

2x6

1x8

TOWING BAR 1/2"x3"x18" FLAT STEEL

ANGLE IRON 7 REQ'D.

2x14

TOP VIEW OF BED FRAME

2x14

3x14

3x10

3x14

2x14

3x10

C OF AXLE

6'-0" ANGLE IRON

2x14

15'-2"

7'-3 1/2"

6'-10"

12 1/2"

2'-8"

6'-3 3/4"

SIDE VIEW WITH THE 2x14 BOARD OFF

8" T HINGES

2x4

2x6

4x4

3x10x9'-0" LONG

DRILL HOLE FOR 5/8" ∅ BOLT

1/2" ∅ CARRIAGE BOLTS

2'-0"

A

A

The Complete Guide to Building Classic Barns, Fences, Storage Sheds, Animal Pens, Outbuildings, Greenhouses, Farm Equipment, & Tools

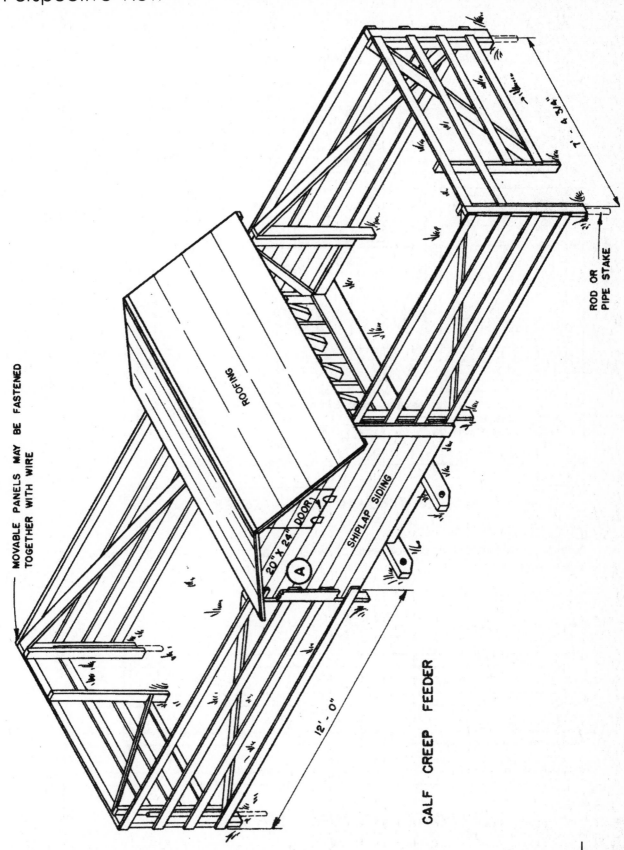

ROOFING

MOVABLE PANELS MAY BE FASTENED
TOGETHER WITH WIRE

SHIPLAP SIDING

20" X 24" DOOR

A

12' - 0"

CALF CREEP FEEDER

7' - 4 3/4"

ROD OR
PIPE STAKE

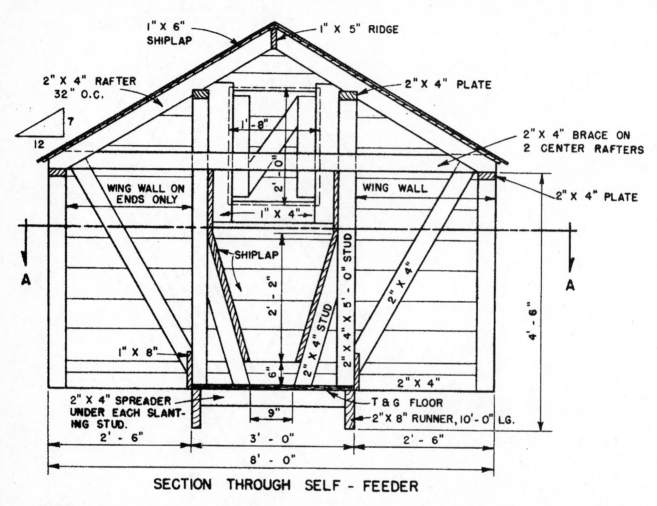

SECTION THROUGH SELF - FEEDER

NOTE

RUNNERS, PANEL POSTS & BOTTOM RAIL OF PANEL TO BE TREATED WITH PRESERVATIVE.
WIRE FENCE PANELS TO OLD ROD OR PIPE STAKES AT FOUR CORNERS.

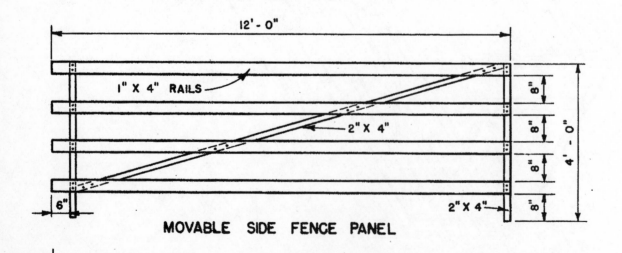

MOVABLE SIDE FENCE PANEL

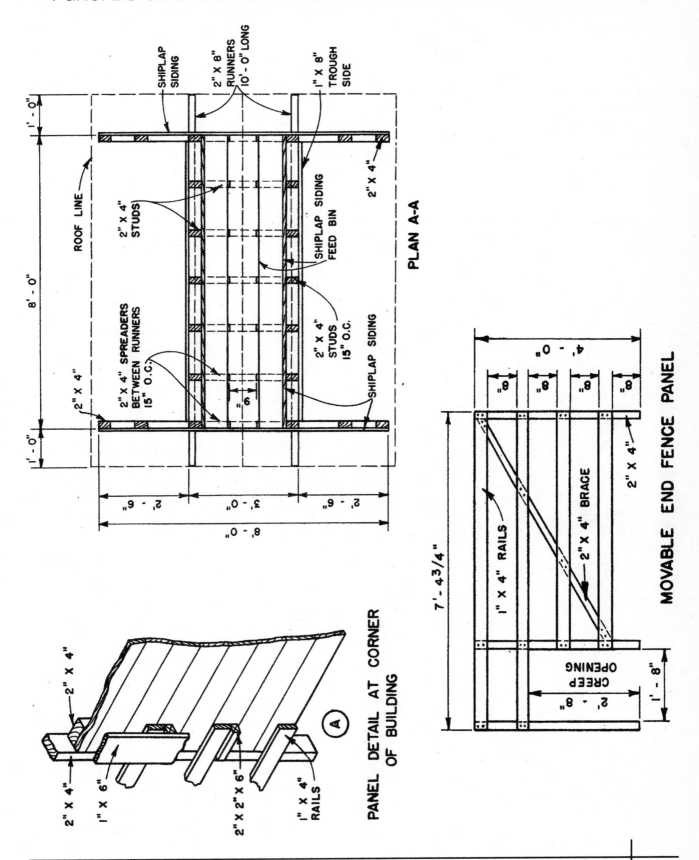

SHIPLAP SIDING

2" X 8" RUNNERS 10'-0" LONG

1" X 8" TROUGH SIDE

ROOF LINE

2" X 4" STUDS

SHIPLAP SIDING FEED BIN

2" X 4"

2" X 4" SPREADERS BETWEEN RUNNERS 15" O.C.

2" X 4" STUDS 15" O.C.

SHIPLAP SIDING

2" X 4"

1'-0"

8'-0"

1'-0"

2'-6"

3'-0"

2'-6"

8'-0"

6"

PLAN A-A

2" X 4"

2" X 4"

1" X 6"

2" X 2" X 6"

1" X 4" RAILS

PANEL DETAIL AT CORNER OF BUILDING

Ⓐ

4'-0"

8" 8" 8" 8"

2" X 4"

7'-4 3/4"

1" X 4" RAILS

2" X 4" BRACE

CREEP OPENING

2'-8"

1'-8"

MOVABLE END FENCE PANEL

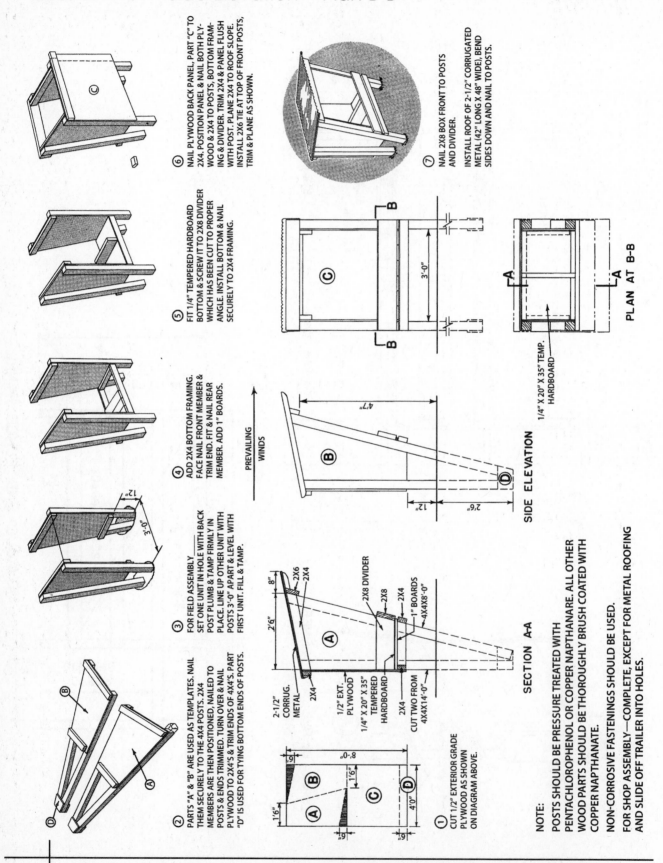

⑥ NAIL PLYWOOD BACK PANEL, PART "C" TO 2X4. POSITION PANEL & NAIL BOTH PLYWOOD & 2X4 TO POSTS, BOTTOM FRAMING & DIVIDER. TRIM 2X4 & PANEL FLUSH WITH POST. PLANE 2X4 TO ROOF SLOPE. INSTALL 2X6 TIE AT TOP OF FRONT POSTS, TRIM & PLANE AS SHOWN.

⑦ NAIL 2X8 BOX FRONT TO POSTS AND DIVIDER.

INSTALL ROOF OF 2-1/2" CORRUGATED METAL (42" LONG X 48" WIDE), BEND SIDES DOWN AND NAIL TO POSTS.

⑤ FIT 1/4" TEMPERED HARDBOARD BOTTOM & SCREW IT TO 2X8 DIVIDER WHICH HAS BEEN CUT TO PROPER ANGLE. INSTALL BOTTOM & NAIL SECURELY TO 2X4 FRAMING.

④ ADD 2X4 BOTTOM FRAMING. FACE NAIL FRONT MEMBER & TRIM END. FIT & NAIL REAR MEMBER. ADD 1" BOARDS.

③ FOR FIELD ASSEMBLY SET ONE UNIT IN HOLE WITH BACK POST PLUMB & TAMP FIRMLY IN PLACE. LINE UP OTHER UNIT WITH POSTS 3'-0" APART & LEVEL WITH FIRST UNIT. FILL & TAMP.

② PARTS "A" & "B" ARE USED AS TEMPLATES. NAIL THEM SECURELY TO THE 4X4 POSTS. 2X4 MEMBERS ARE THEN POSITIONED, NAILED TO POSTS & ENDS TRIMMED. TURN OVER & NAIL PLYWOOD TO 2X4'S & TRIM ENDS OF 4X4'S. PART "D" IS USED FOR TYING BOTTOM ENDS OF POSTS.

① CUT 1/2" EXTERIOR GRADE PLYWOOD AS SHOWN ON DIAGRAM ABOVE.

PLAN AT B-B

SIDE ELEVATION

PREVAILING WINDS

SECTION A-A

2-1/2" CORRUG. METAL

1/2" EXT. PLYWOOD

1/4" X 20" X 35" TEMPERED HARDBOARD

2X4 CUT TWO FROM 4X4X14'-0"

2X6
2X4
2X8 DIVIDER
2X8
2X4
1" BOARDS
4X4X8'-0"

1/4" X 20" X 35" TEMP. HARDBOARD

NOTE:

POSTS SHOULD BE PRESSURE TREATED WITH PENTACHLOROPHENOL OR COPPER NAPTHANARE. ALL OTHER WOOD PARTS SHOULD BE THOROUGHLY BRUSH COATED WITH COPPER NAPTHANATE.

NON-CORROSIVE FASTENINGS SHOULD BE USED.

FOR SHOP ASSEMBLY—COMPLETE, EXCEPT FOR METAL ROOFING AND SLIDE OFF TRAILER INTO HOLES.

• Isometric View

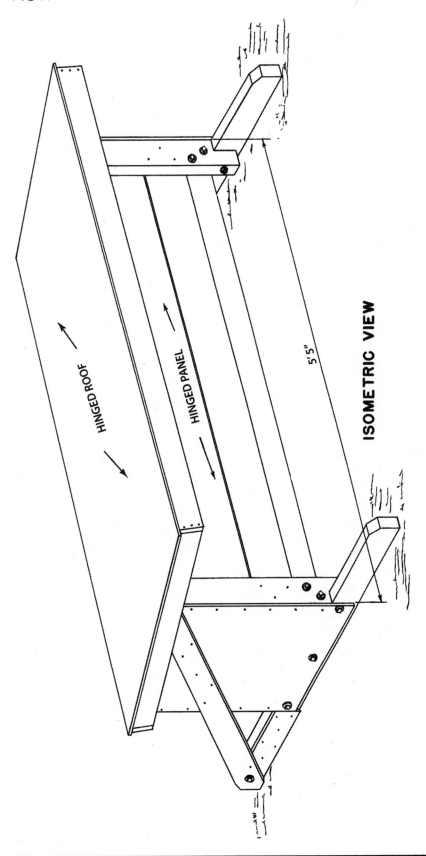

HINGED ROOF

HINGED PANEL

5'5"

ISOMETRIC VIEW

• Cutting Diagram

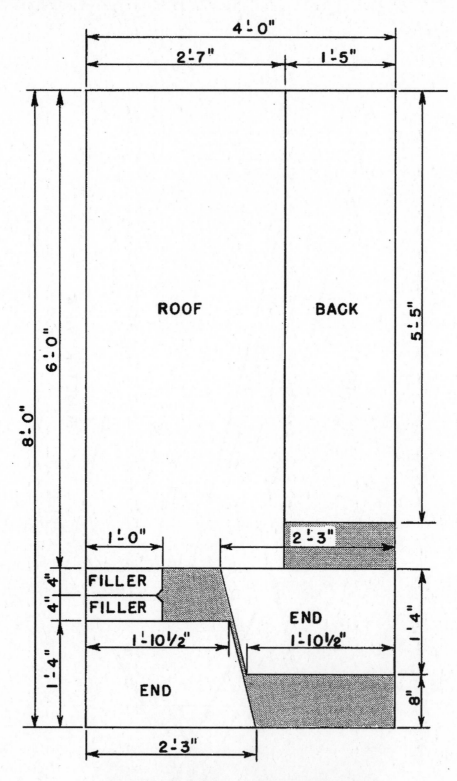

CUTTING DIAGRAM
(3/8" EXTERIOR TYPE PLYWOOD)

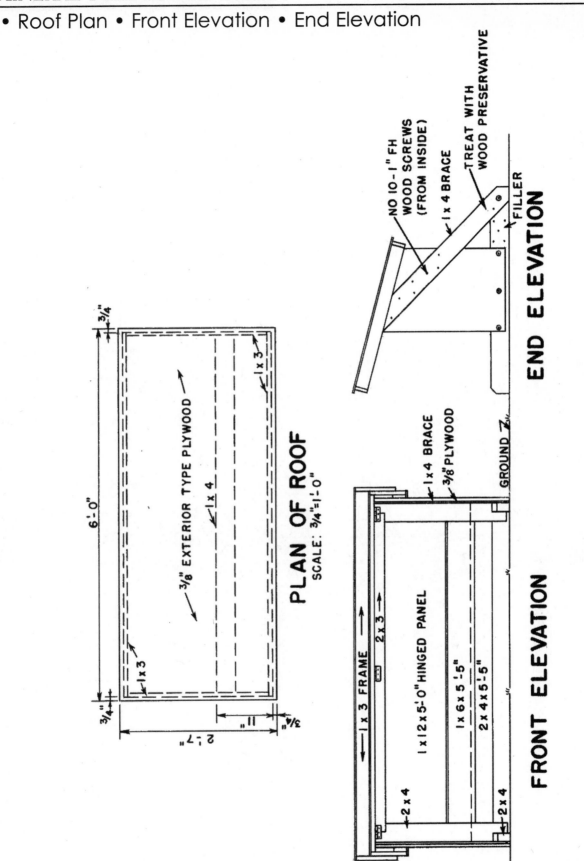

PLAN OF ROOF
SCALE: 3/4"=1'-0"

END ELEVATION

FRONT ELEVATION

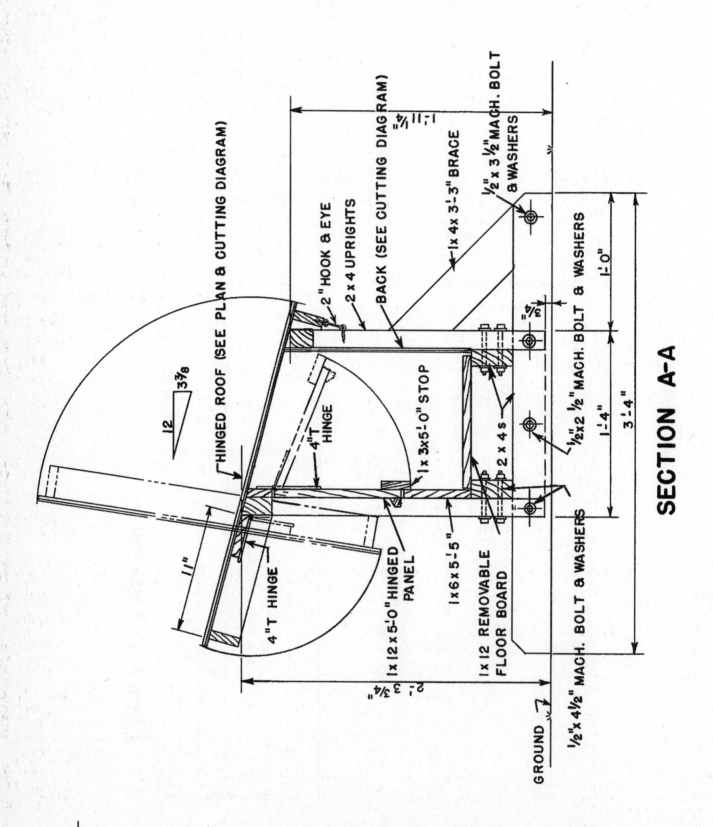

• Front View • Perspective View

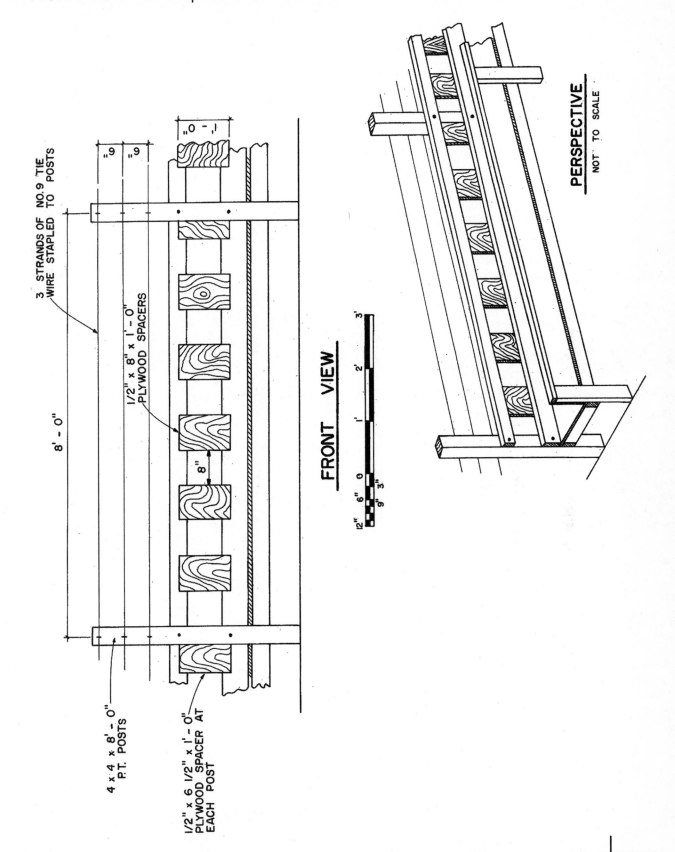

3 STRANDS OF NO.9 TIE WIRE STAPLED TO POSTS

8' - 0"

1/2" x 8" x 1'-0" PLYWOOD SPACERS

8"

4 x 4 x 8'-0" P.T. POSTS

1/2" x 6 1/2" x 1'-0" PLYWOOD SPACER AT EACH POST

FRONT VIEW

PERSPECTIVE

NOT TO SCALE

• Cross Section

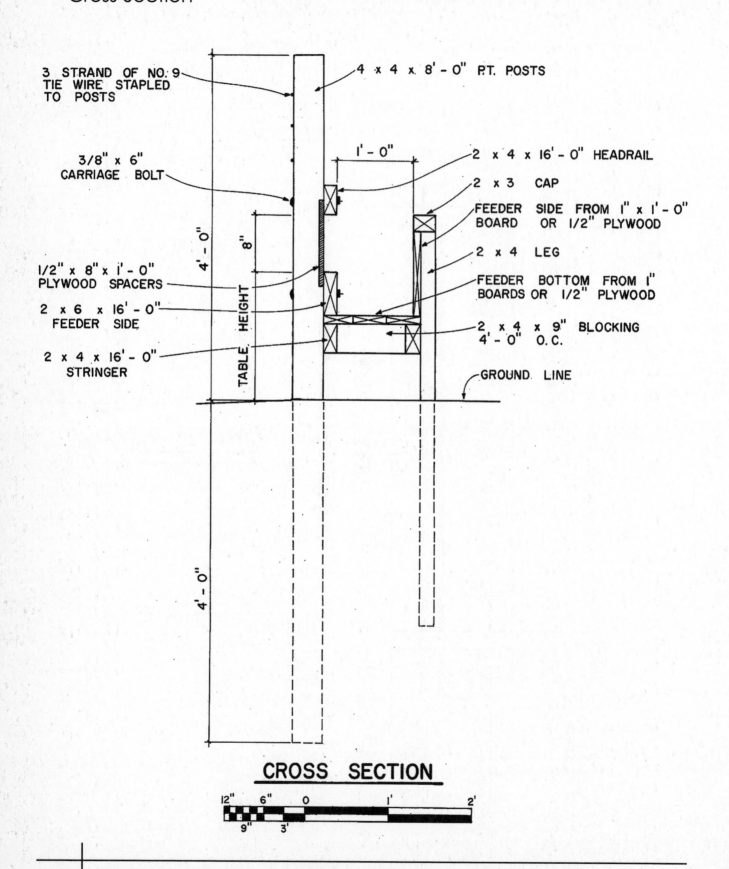

3 STRAND OF NO. 9
TIE WIRE STAPLED
TO POSTS

4 x 4 x 8' - 0" P.T. POSTS

1' - 0"

3/8" x 6"
CARRIAGE BOLT

2 x 4 x 16' - 0" HEADRAIL

2 x 3 CAP

FEEDER SIDE FROM 1" x 1' - 0"
BOARD OR 1/2" PLYWOOD

2 x 4 LEG

1/2" x 8" x 1' - 0"
PLYWOOD SPACERS

FEEDER BOTTOM FROM 1"
BOARDS OR 1/2" PLYWOOD

2 x 6 x 16' - 0"
FEEDER SIDE

2 x 4 x 9" BLOCKING
4' - 0" O.C.

2 x 4 x 16' - 0"
STRINGER

GROUND LINE

4' - 0"

8"

TABLE HEIGHT

4' - 0"

CROSS SECTION

12"   6"   0   1'   2'
9"   3'

• Table of Feeder Height • Optional Silage Deflector

TABLE OF FEEDER HEIGHT

SHEEP SIZE	HEIGHT TO THROAT 'H'
SMALL BREEDS & FEEDER LAMBS	10" TO 1'-6"
MEDIUM BREEDS	1'-5" TO 1'-8"
LARGE BREEDS	1'-8" TO 1'-10"

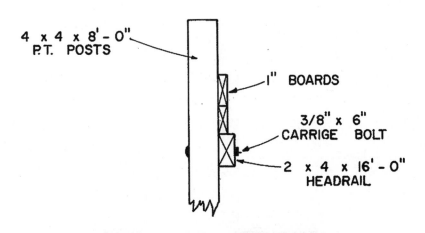

4 x 4 x 8'-0"
P.T. POSTS

1" BOARDS

3/8" x 6"
CARRIGE BOLT

2 x 4 x 16'-0"
HEADRAIL

DETAIL OF OPTIONAL SILAGE DEFLECTOR

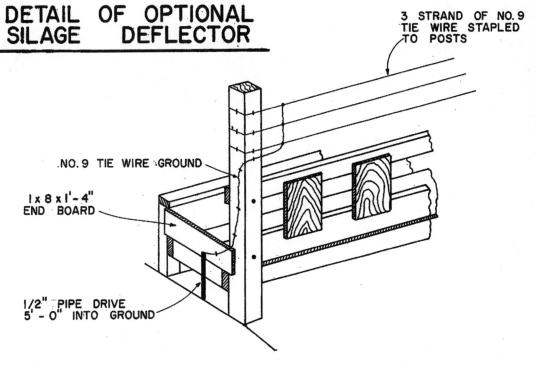

3 STRAND OF NO. 9
TIE WIRE STAPLED
TO POSTS

NO. 9 TIE WIRE GROUND

1 x 8 x 1'-4"
END BOARD

1/2" PIPE DRIVE
5'-0" INTO GROUND

END VIEW
NOT TO SCALE

• Plan • Perspective View

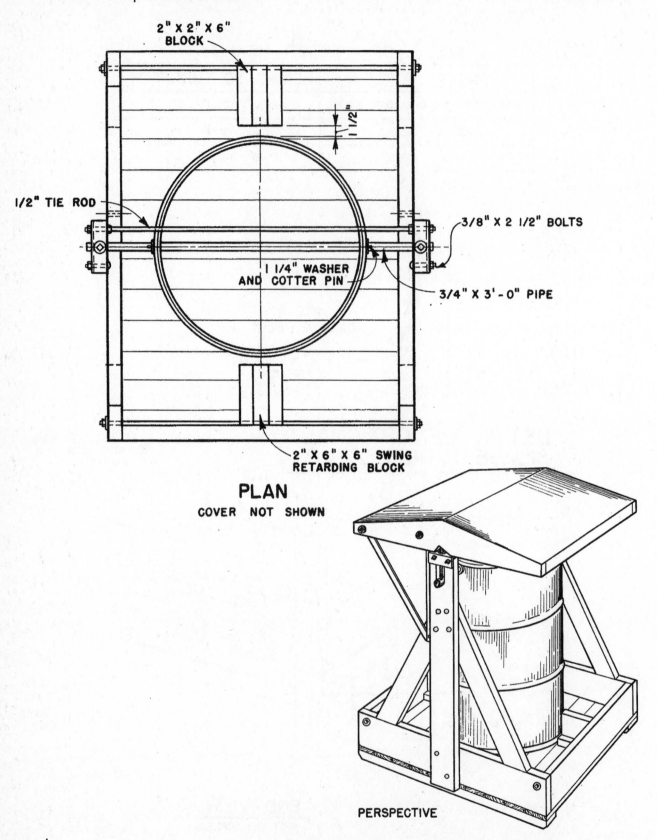

2" X 2" X 6"
BLOCK

1 1/2"

1/2" TIE ROD

3/8" X 2 1/2" BOLTS

1 1/4" WASHER
AND COTTER PIN

3/4" X 3' - 0" PIPE

2" X 6" X 6" SWING
RETARDING BLOCK

PLAN
COVER NOT SHOWN

PERSPECTIVE

• Framing Details

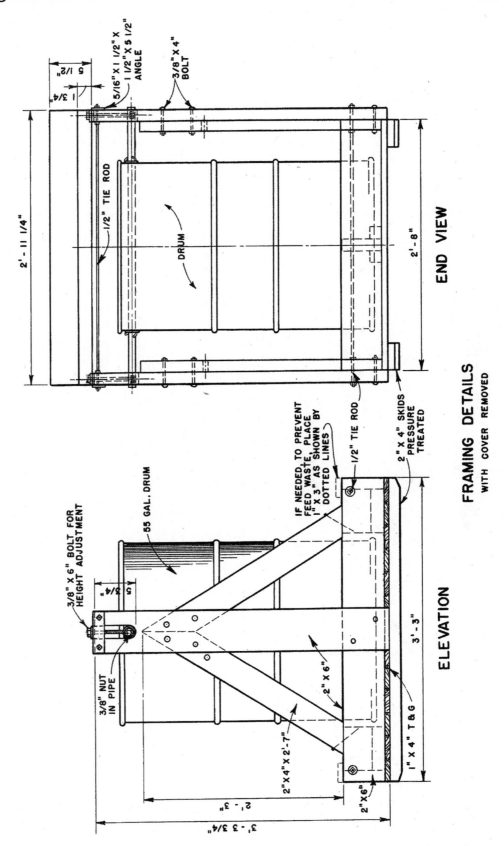

END VIEW

FRAMING DETAILS
WITH COVER REMOVED

ELEVATION

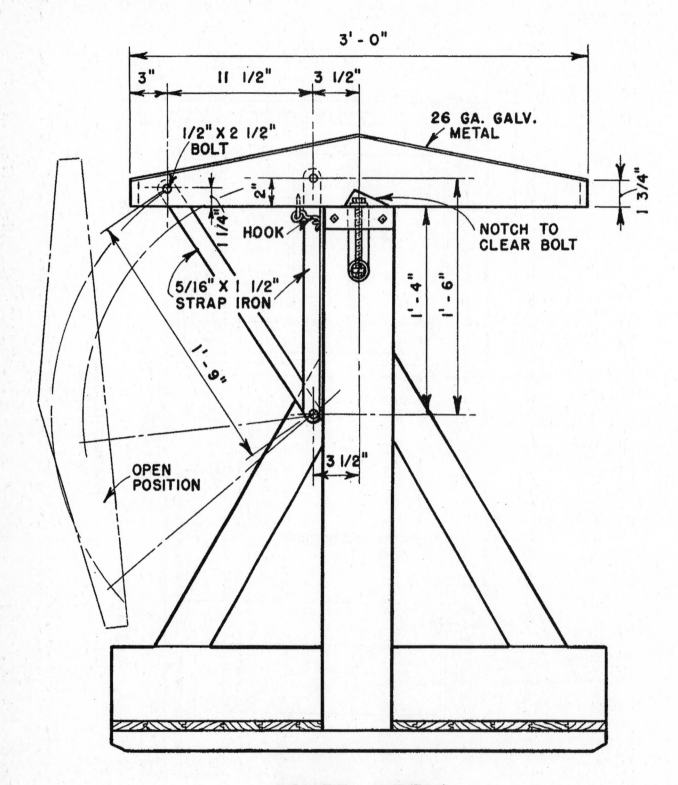

COVER DETAIL
SHOWING FIT

chapter 8

FENCES & GATES

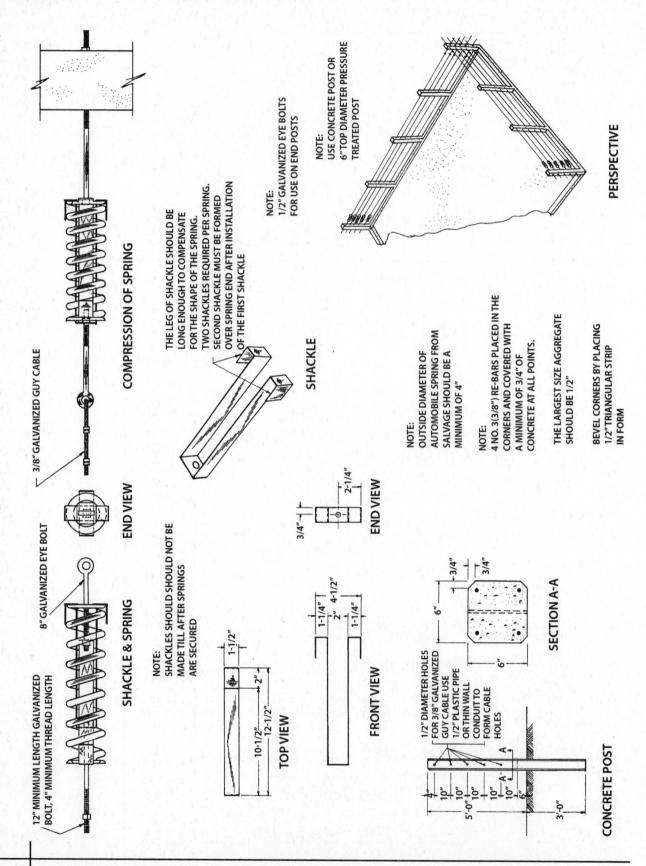

3/8" GALVANIZED GUY CABLE

COMPRESSION OF SPRING

8" GALVANIZED EYE BOLT

END VIEW

12" MINIMUM LENGTH GALVANIZED BOLT, 4" MINIMUM THREAD LENGTH

SHACKLE & SPRING

NOTE:
SHACKLES SHOULD SHOULD NOT BE MADE TILL AFTER SPRINGS ARE SECURED

THE LEG OF SHACKLE SHOULD BE LONG ENOUGH TO COMPENSATE FOR THE SHAPE OF THE SPRING. TWO SHACKLES REQUIRED PER SPRING. SECOND SHACKLE MUST BE FORMED OVER SPRING END AFTER INSTALLATION OF THE FIRST SHACKLE

SHACKLE

NOTE:
1/2" GALVANIZED EYE BOLTS FOR USE ON END POSTS

NOTE:
USE CONCRETE POST OR 6" TOP DIAMETER PRESSURE TREATED POST

PERSPECTIVE

2-1/4"
3/4"

END VIEW

1-1/2"
2"
10-1/2"
12-1/2"

TOP VIEW

1-1/4"
4-1/2"
2"
1-1/4"

FRONT VIEW

NOTE:
OUTSIDE DIAMETER OF AUTOMOBILE SPRING FROM SALVAGE SHOULD BE A MINIMUM OF 4"

NOTE:
4 NO. 3 (3/8") RE-BARS PLACED IN THE CORNERS AND COVERED WITH A MINIMUM OF 3/4" OF CONCRETE AT ALL POINTS.

THE LARGEST SIZE AGGREGATE SHOULD BE 1/2"

BEVEL CORNERS BY PLACING 1/2" TRIANGULAR STRIP IN FORM

3/4"
3/4"
6"
6"

SECTION A-A

1/2" DIAMETER HOLES FOR 3/8" GALVANIZED GUY CABLE USE 1/2" PLASTIC PIPE OR THIN WALL CONDUIT TO FORM CABLE HOLES

A
A

4"
10"
10"
10"
10"
10"
6"
5'-0"
3'-0"

CONCRETE POST

WOODEN FENCE TYPES I

- Alternate Board Fence • Novelty Pattern • Louvered Fence
- Basket Weave Fence

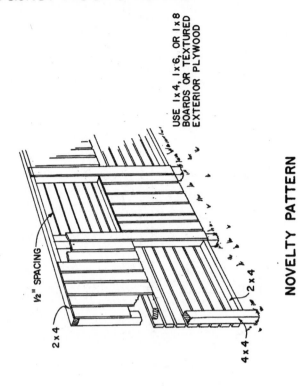

NOVELTY PATTERN

USE 1x4, 1x6, OR 1x8 BOARDS OR TEXTURED EXTERIOR PLYWOOD

½" SPACING

2 x 4

2 x 4

4 x 4

BASKET WEAVE FENCE

½" x 6" BOARDS

1 x 4 CAP

1 x 2

4 x 4 POSTS

1 x 2 NAILING STRIP

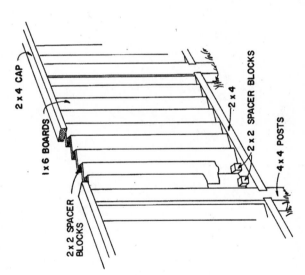

LOUVERED FENCE

2 x 4 CAP

2 x 4

2 x 2 SPACER BLOCKS

1 x 6 BOARDS

2 x 2 SPACER BLOCKS

4 x 4 POSTS

ALTERNATE BOARD FENCE

4 x 4

2 x 4 SUPPORTS

1 x 6 CAP OPTIONAL

1 x 6 BOARDS STAGGERED

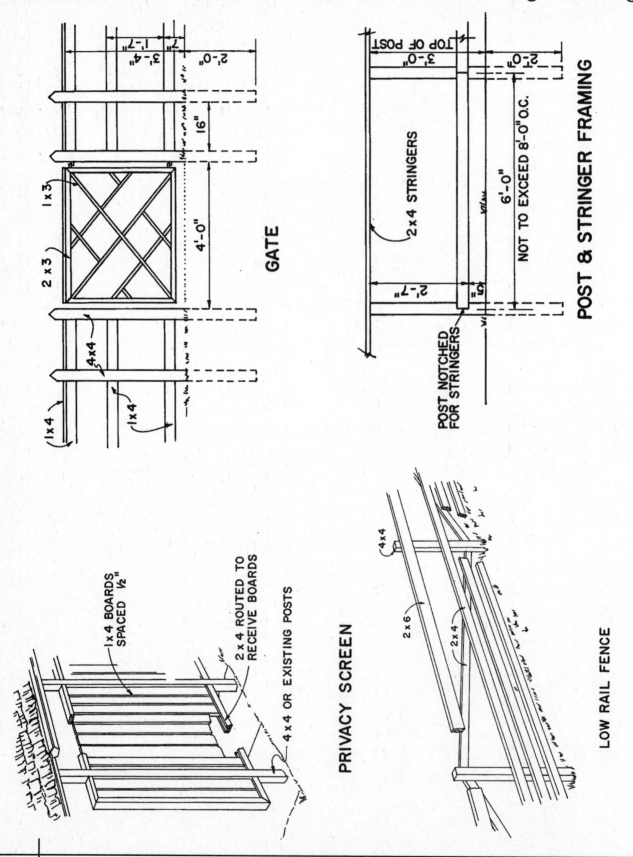

GATE

POST & STRINGER FRAMING

PRIVACY SCREEN

LOW RAIL FENCE

WOODEN FENCE TYPES III

• Louvered Gate • Stretcher Fence

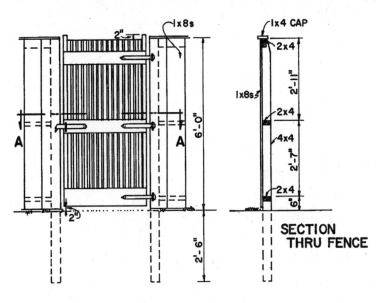

LOUVERED GATE

SECTION THRU FENCE

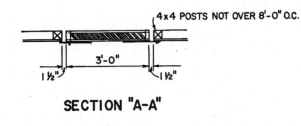

SECTION "A-A"

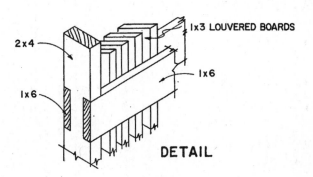

DETAIL

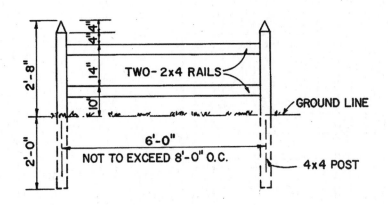

STRETCHER FENCE

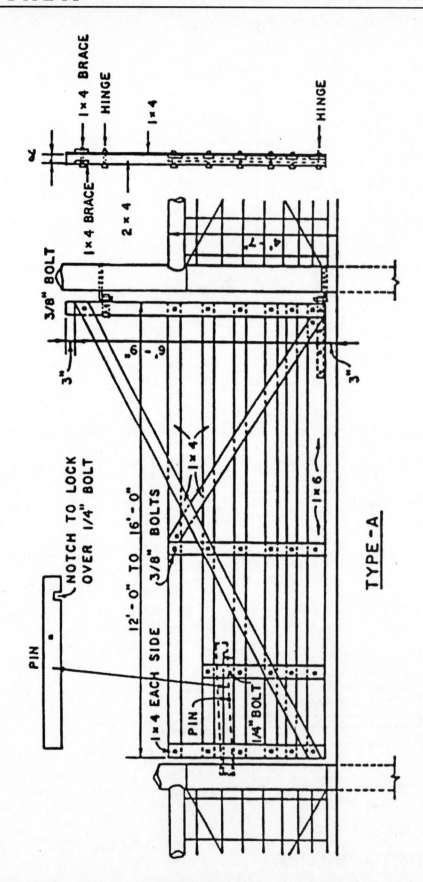

TYPE-A

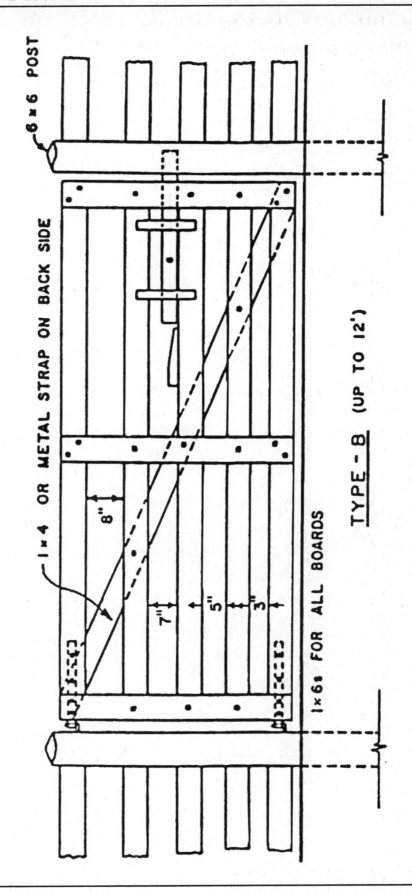

TYPE - B (UP TO 12')

6×6 POST

1×4 OR METAL STRAP ON BACK SIDE

1×6s FOR ALL BOARDS

8"

7"

5"

3"

• Steps Over Fence • Childproof Latch • Fence Passage

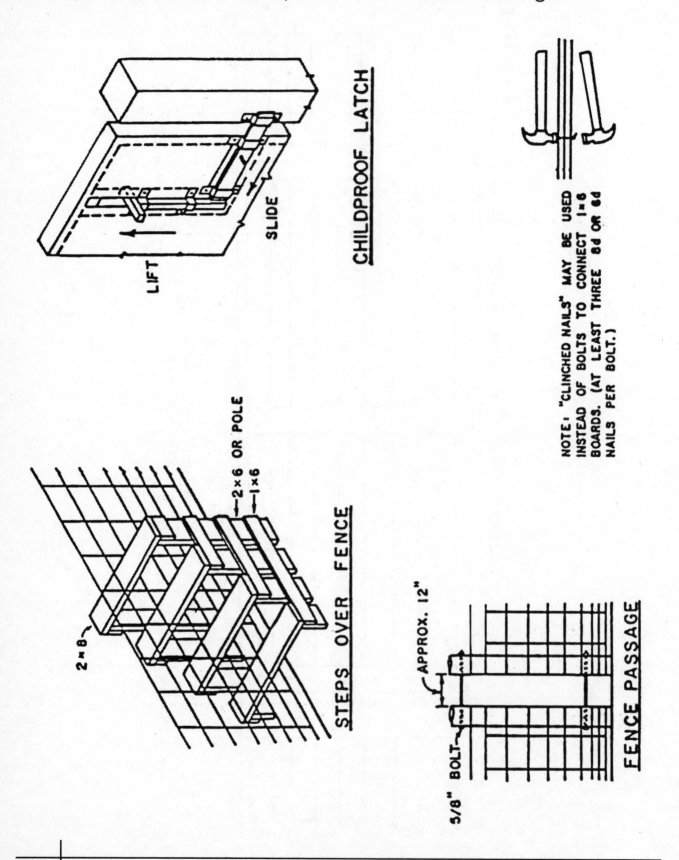

CHILDPROOF LATCH

LIFT

SLIDE

NOTE: "CLINCHED NAILS" MAY BE USED INSTEAD OF BOLTS TO CONNECT 1×6 BOARDS. (AT LEAST THREE 8d OR 6d NAILS PER BOLT.)

2×6 OR POLE

1×6

2×8

STEPS OVER FENCE

APPROX. 12"

5/8" BOLT

FENCE PASSAGE

• Plans

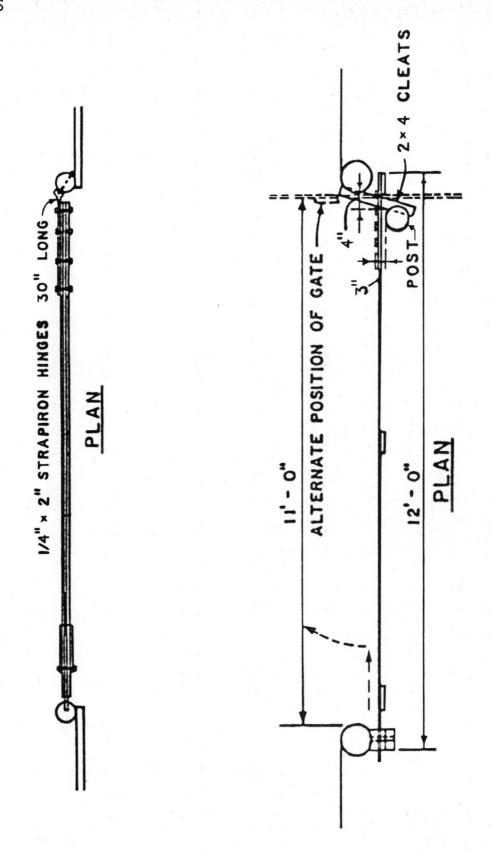

1/4" × 2" STRAPIRON HINGES 30" LONG

PLAN

2 × 4 CLEATS

ALTERNATE POSITION OF GATE

11' - 0"

12' - 0"

4"

3"

POST

PLAN

• Plan

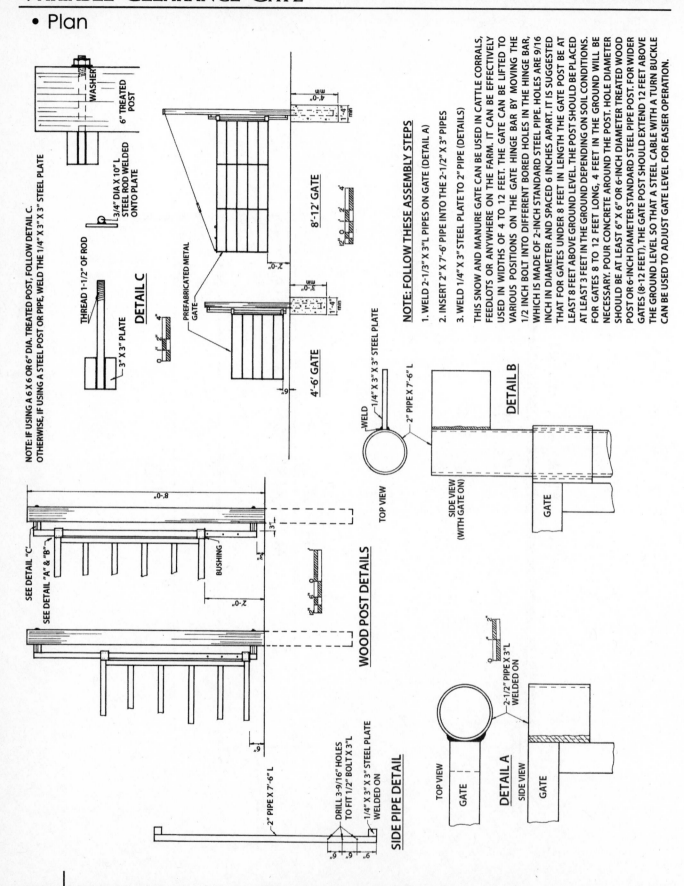

NOTE: FOLLOW THESE ASSEMBLY STEPS

1. WELD 2-1/3" X 3" L PIPES ON GATE (DETAIL A)

2. INSERT 2" X 7'-6' PIPE INTO THE 2-1/2" X 3" PIPES

3. WELD 1/4" X 3" STEEL PLATE TO 2" PIPE (DETAILS)

THIS SNOW AND MANURE GATE CAN BE USED IN CATTLE CORRALS, FEEDLOTS OR ANYWHERE ON THE FARM. IT CAN BE EFFECTIVELY USED IN WIDTHS OF 4 TO 12 FEET. THE GATE CAN BE LIFTED TO VARIOUS POSITIONS ON THE GATE HINGE BAR BY MOVING THE 1/2 INCH BOLT INTO DIFFERENT BORED HOLES IN THE HINGE BAR, WHICH IS MADE OF 2-INCH STANDARD STEEL PIPE. HOLES ARE 9/16 INCH IN DIAMETER AND SPACED 6 INCHES APART. IT IS SUGGESTED THAT FOR GATES UNDER 8 FEET IN LENGTH THE GATE POST BE AT LEAST 8 FEET ABOVE GROUND LEVEL. THE POST SHOULD BE PLACED AT LEAST 3 FEET IN THE GROUND DEPENDING ON SOIL CONDITIONS. FOR GATES 8 TO 12 FEET LONG, 4 FEET IN THE GROUND WILL BE NECESSARY. POUR CONCRETE AROUND THE POST. HOLE DIAMETER SHOULD BE AT LEAST 6" X 6" OR 6-INCH DIAMETER TREATED WOOD POST OR 6-INCH DIAMETER STANDARD STEEL PIPE POST. FOR WIDER GATES (8-12 FEET), THE GATE POST SHOULD EXTEND 12 FEET ABOVE THE GROUND LEVEL SO THAT A STEEL CABLE WITH A TURN BUCKLE CAN BE USED TO ADJUST GATE LEVEL FOR EASIER OPERATION.

Automatic Gate
• Perspective View

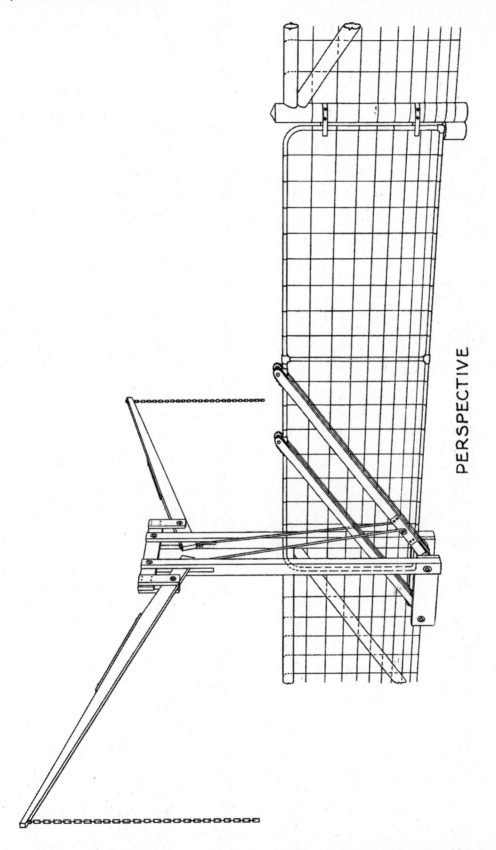

PERSPECTIVE

• Side Elevation

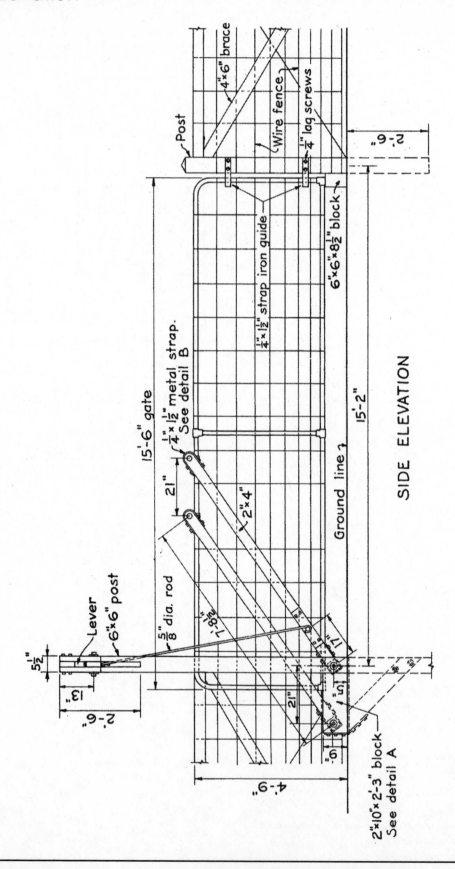

• End Elevation

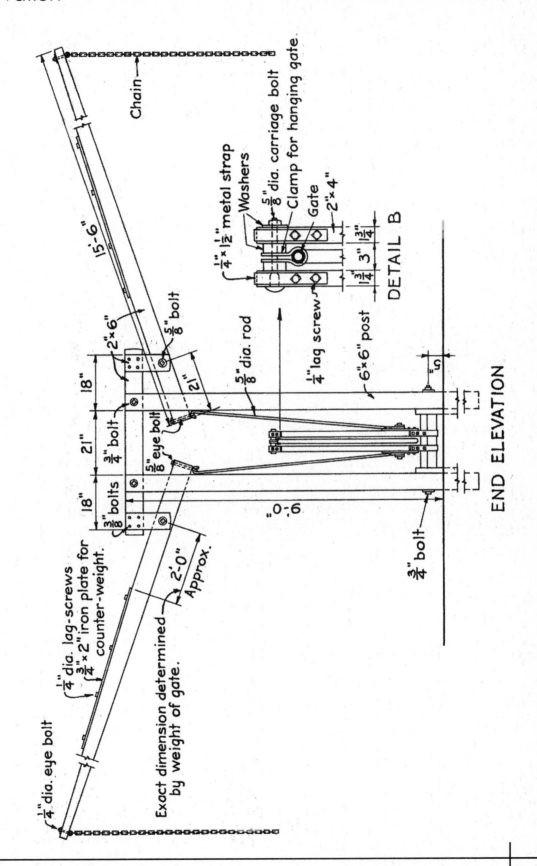

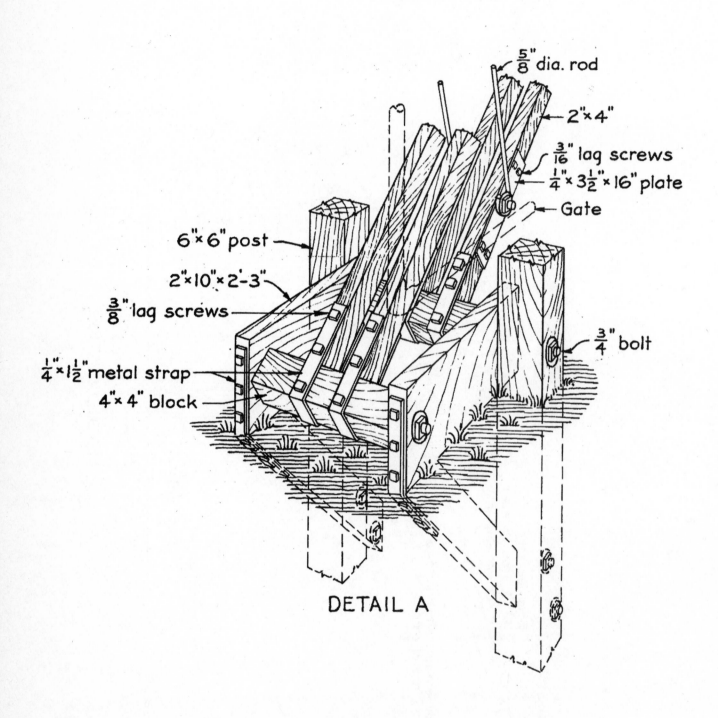

$\frac{5}{8}$" dia. rod

2"× 4"

$\frac{3}{16}$" lag screws

$\frac{1}{4}$"× 3$\frac{1}{2}$"× 16" plate

Gate

6"× 6" post

2"× 10"× 2'-3"

$\frac{3}{8}$" lag screws

$\frac{3}{4}$" bolt

$\frac{1}{4}$"× 1$\frac{1}{2}$" metal strap

4"× 4" block

DETAIL A

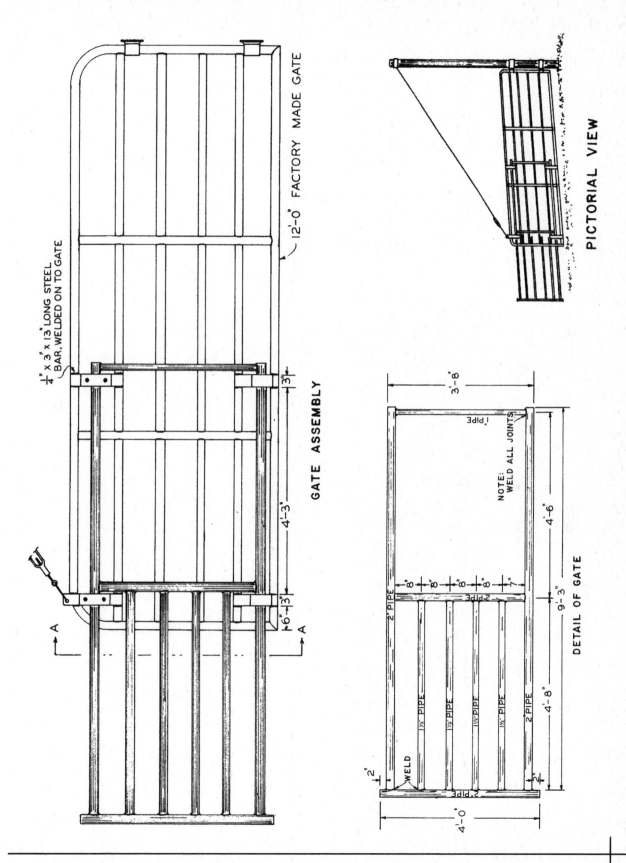

PICTORIAL VIEW

GATE ASSEMBLY

$\frac{1}{4}$" X 3" X 13" LONG STEEL BAR, WELDED ON TO GATE

12'-0" FACTORY MADE GATE

3"

4'-3"

6"—3"

A

A

DETAIL OF GATE

3'-8"

1" PIPE

NOTE: WELD ALL JOINTS

4'-6"

9'-3"

2" PIPE

3" PIPE

2" PIPE

8" 8" 8" 8" 7"

1½" PIPE

4'-8"

2" PIPE

WELD

2"

2"

2" PIPE

4'-0"

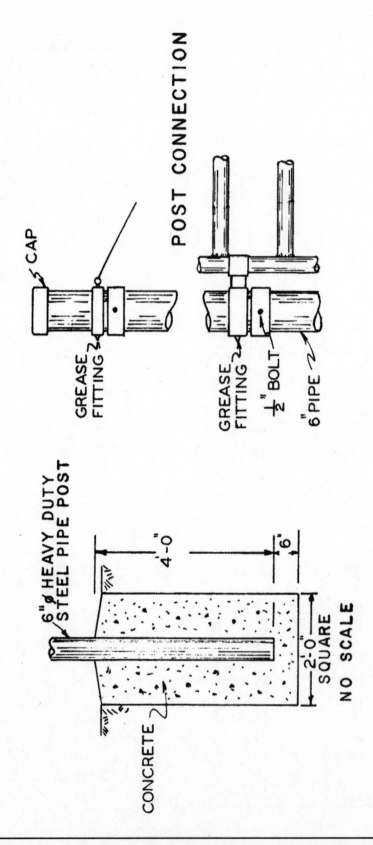

POST CONNECTION

CAP

GREASE FITTING

GREASE FITTING

½" BOLT

6" PIPE

6"ø HEAVY DUTY STEEL PIPE POST

4'-0"

6"

2'-0" SQUARE

NO SCALE

CONCRETE

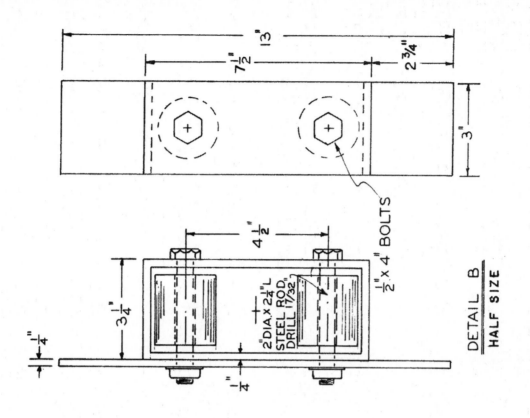

DETAIL B
HALF SIZE

$\frac{1}{2}$" x 4" BOLTS

2"DIA.x 2$\frac{1}{4}$"L
STEEL ROD
DRILL 17/32"

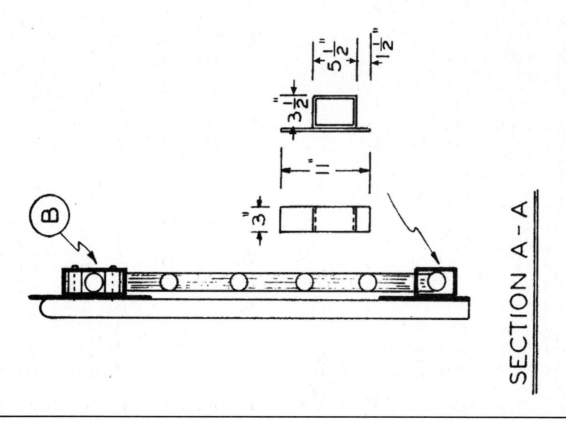

SECTION A-A

• Perspective View

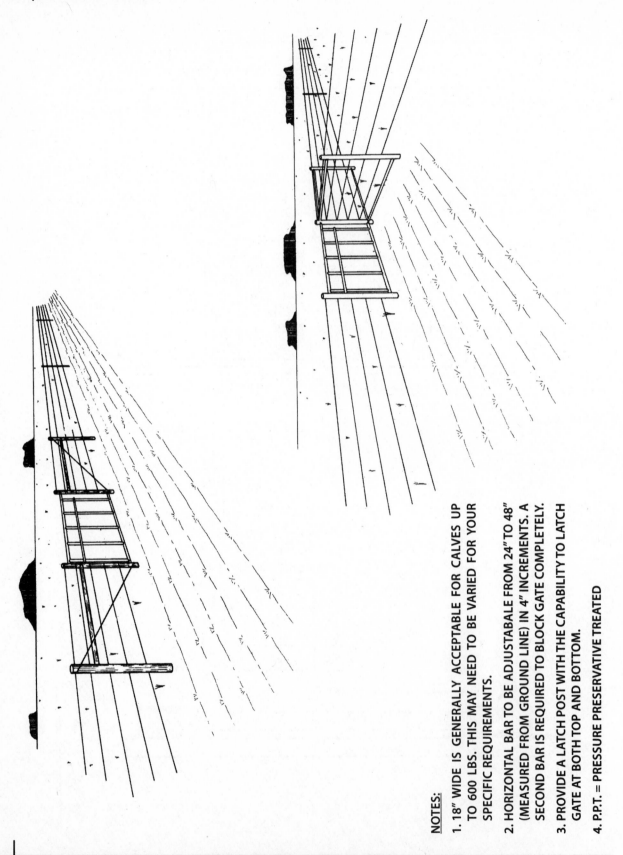

NOTES:

1. 18" WIDE IS GENERALLY ACCEPTABLE FOR CALVES UP TO 600 LBS. THIS MAY NEED TO BE VARIED FOR YOUR SPECIFIC REQUIREMENTS.

2. HORIZONTAL BAR TO BE ADJUSTABALE FROM 24" TO 48" (MEASURED FROM GROUND LINE) IN 4" INCREMENTS. A SECOND BAR IS REQUIRED TO BLOCK GATE COMPLETELY.

3. PROVIDE A LATCH POST WITH THE CAPABILITY TO LATCH GATE AT BOTH TOP AND BOTTOM.

4. P.P.T. = PRESSURE PRESERVATIVE TREATED

• Metal & Wood Construction Stretch Post Assembly

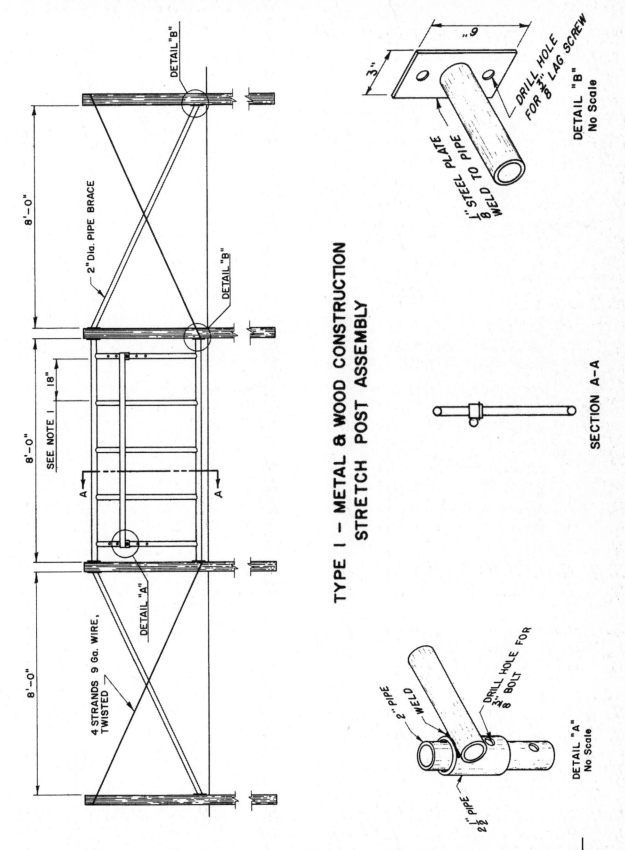

DETAIL "B"

DETAIL "B"

8'-0"

2" Dia. PIPE BRACE

SEE NOTE 1

18"

8'-0"

A   A

4 STRANDS 9 Ga. WIRE, TWISTED

DETAIL "A"

8'-0"

TYPE I – METAL & WOOD CONSTRUCTION
STRETCH POST ASSEMBLY

9"

3"

⅛" STEEL PLATE
WELD TO PIPE

DRILL HOLE FOR ⅜" LAG SCREW

DETAIL "B"
No Scale

SECTION A-A

2" PIPE

WELD

DRILL HOLE FOR
½"ø BOLT

2½" PIPE

DETAIL "A"
No Scale

• Wood Construction Stretch Post Assembly

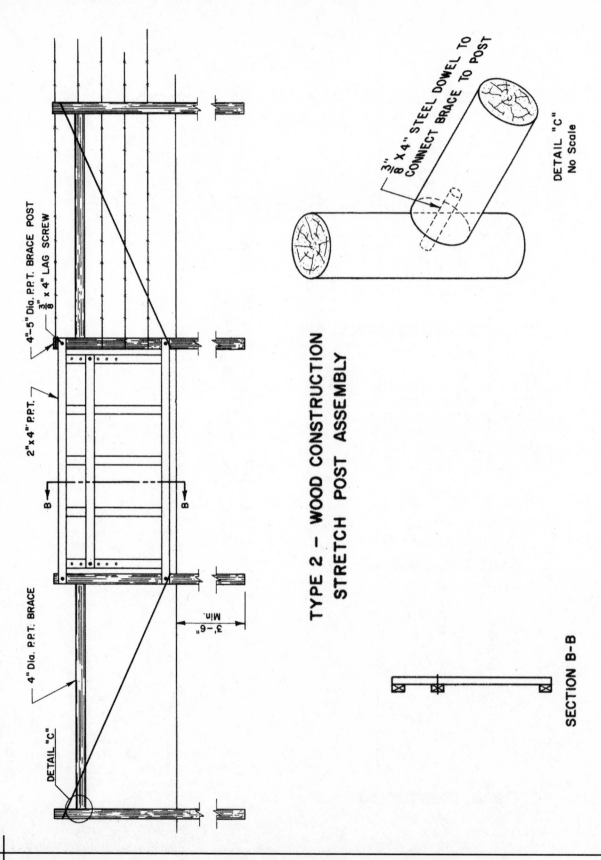

4"–5" Dia. P.P.T. BRACE POST

$\frac{3}{8}$" x 4" LAG SCREW

2"x 4" P.P.T.

B

4" Dia. P.P.T. BRACE

DETAIL "C"

3'–6" Min.

$\frac{3}{8}$" X 4" STEEL DOWEL TO CONNECT BRACE TO POST

DETAIL "C"
No Scale

TYPE 2 – WOOD CONSTRUCTION
STRETCH POST ASSEMBLY

SECTION B-B

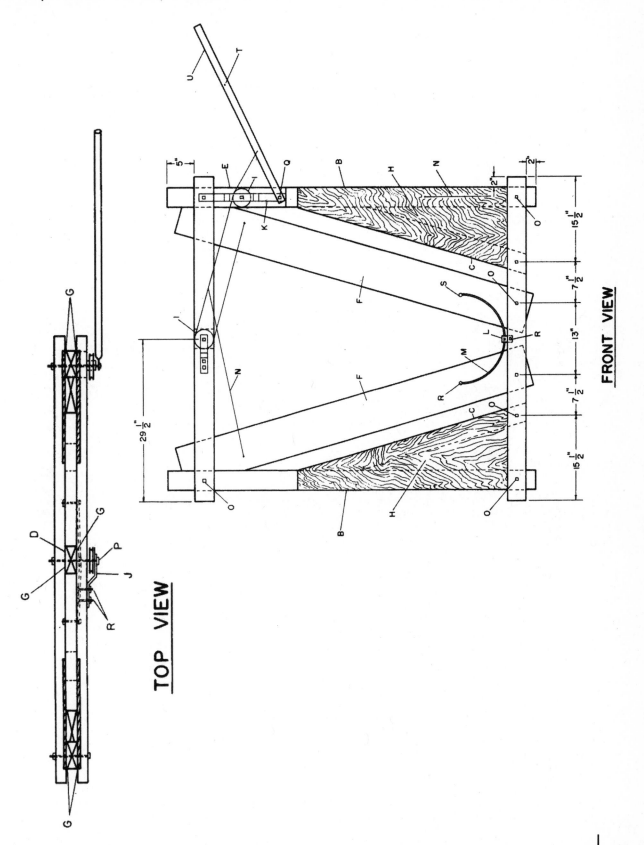

FRONT VIEW

TOP VIEW

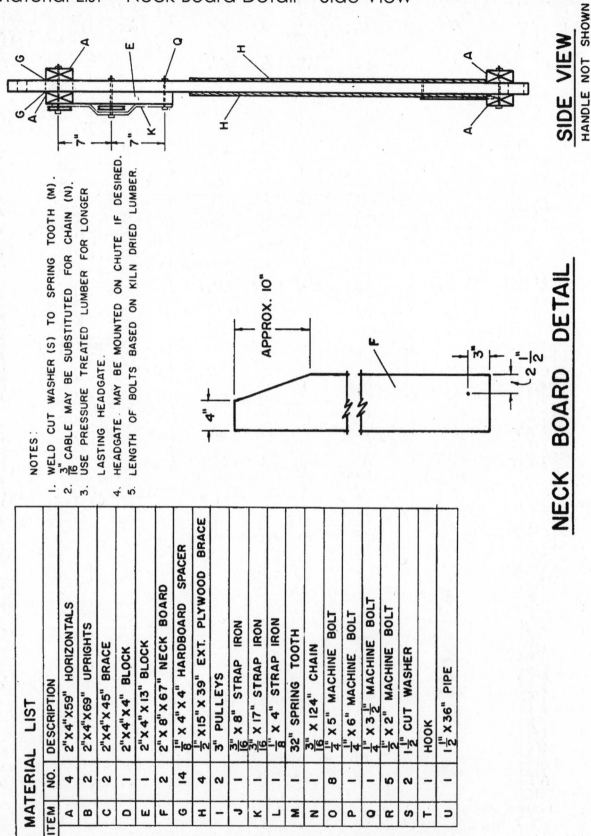

SIDE VIEW
HANDLE NOT SHOWN

NECK BOARD DETAIL

NOTES:
1. WELD CUT WASHER (S) TO SPRING TOOTH (M).
2. $\frac{3}{16}$" CABLE MAY BE SUBSTITUTED FOR CHAIN (N).
3. USE PRESSURE TREATED LUMBER FOR LONGER LASTING HEADGATE.
4. HEADGATE MAY BE MOUNTED ON CHUTE IF DESIRED.
5. LENGTH OF BOLTS BASED ON KILN DRIED LUMBER.

MATERIAL LIST

ITEM	NO.	DESCRIPTION
A	4	2"X4"X59" HORIZONTALS
B	2	2"X4"X69" UPRIGHTS
C	2	2"X4"X45" BRACE
D	1	2"X4"X4" BLOCK
E	1	2"X4"X13" BLOCK
F	2	2"X8"X67" NECK BOARD
G	14	$\frac{1}{8}$"X4"X4" HARDBOARD SPACER
H	4	$\frac{1}{2}$"X15"X39" EXT. PLYWOOD BRACE
I	2	3" PULLEYS
J	1	$\frac{3}{16}$" X 8" STRAP IRON
K	1	$\frac{3}{16}$" X 17" STRAP IRON
L	1	$\frac{1}{8}$" X 4" STRAP IRON
M	1	32" SPRING TOOTH
N	1	$\frac{3}{16}$" X 124" CHAIN
O	8	$\frac{1}{4}$" X 5" MACHINE BOLT
P	1	$\frac{1}{4}$" X 6" MACHINE BOLT
Q	1	$\frac{1}{4}$" X 3$\frac{1}{2}$" MACHINE BOLT
R	5	$\frac{1}{2}$" X 2" MACHINE BOLT
S	2	1$\frac{1}{2}$" CUT WASHER
T	1	HOOK
U	1	1$\frac{1}{2}$" X 36" PIPE

146

THE COMPLETE GUIDE TO BUILDING CLASSIC BARNS, FENCES, STORAGE SHEDS, ANIMAL PENS, OUTBUILDINGS, GREENHOUSES, FARM EQUIPMENT, & TOOLS

chapter 9

CORRALS & CHUTES

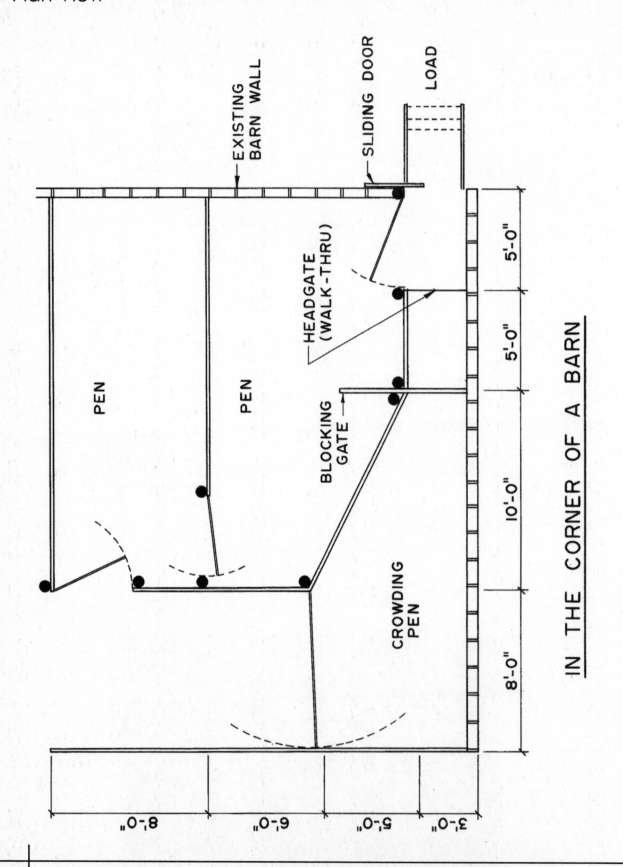

EXISTING BARN WALL

SLIDING DOOR

LOAD

HEADGATE (WALK-THRU)

BLOCKING GATE

PEN

PEN

CROWDING PEN

5'-0"

5'-0"

10'-0"

8'-0"

IN THE CORNER OF A BARN

8'-0"

6'-0"

5'-0"

3'-0"

• Plan View

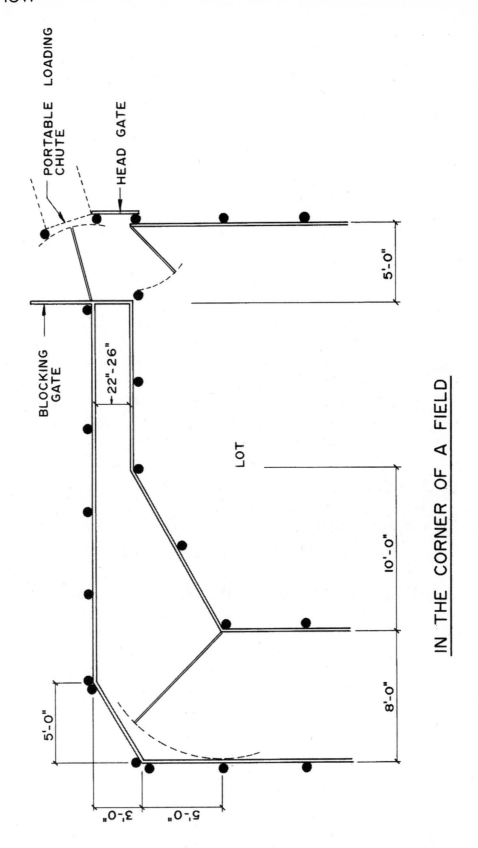

PORTABLE LOADING CHUTE

HEAD GATE

BLOCKING GATE

22"- 26"

LOT

5'-0"

10'-0"

8'-0"

5'-0"

3'-0"

5'-0"

IN THE CORNER OF A FIELD

CONSTRUCTION NOTES & SPECIFICATIONS

HOLDING PENS:
 20 SQ. FT. PER MATURE ANIMAL
 TWO (OR MORE) PENS
 WATER (IF CATTLE ARE
  HELD OVERNIGHT)

CROWDING PEN:
 150 SQ. FT. (ONE TRUCKLOAD)

WORKING CHUTE:
 18 TO 30 FT. LONG
 22" TO 26" WIDE
 BLOCKING GATES TO SPEED
  HANDLING & PREVENT CROWDING
  AT HEADGATE OR SCALES

ADDITIONAL FEATURES:
 PORTABLE SCALES
 SPRAY PEN
 CUTTING GATES FOR SORTING
 SQUEEZE
 WALKING PLATFORM ON ONE OR
  BOTH SIDES OF THE CHUTE

50 Head Beef/Cattle Corral

• Perspective View

Perspective Sketch

• Plan

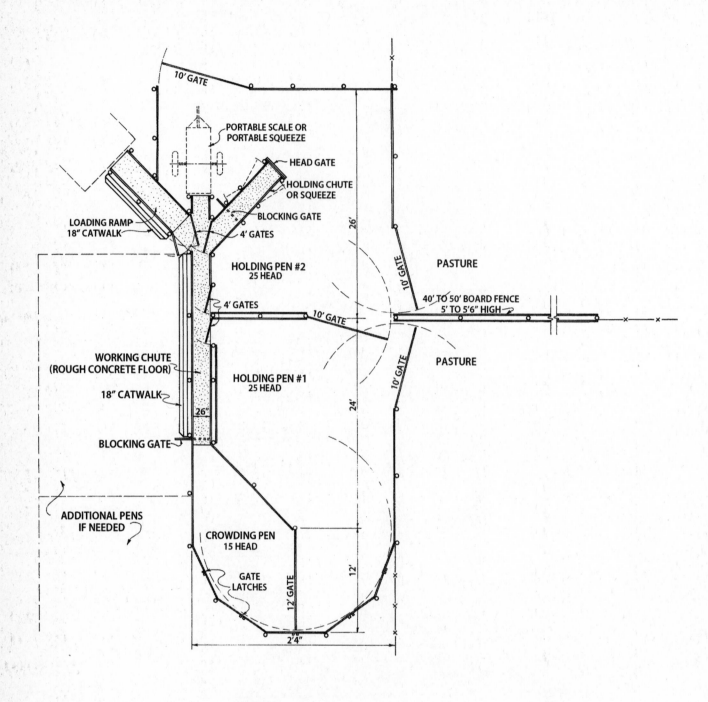

PORTABLE SCALE OR PORTABLE SQUEEZE

HEAD GATE

HOLDING CHUTE OR SQUEEZE

BLOCKING GATE

4' GATES

LOADING RAMP 18" CATWALK

HOLDING PEN #2 25 HEAD

4' GATES

WORKING CHUTE (ROUGH CONCRETE FLOOR)

18" CATWALK

HOLDING PEN #1 25 HEAD

BLOCKING GATE

26"

ADDITIONAL PENS IF NEEDED

CROWDING PEN 15 HEAD

GATE LATCHES

10' GATE

26'

10' GATE

PASTURE

40' TO 50' BOARD FENCE 5' TO 5'6" HIGH

10' GATE

PASTURE

24'

12'

12' GATE

2'4"

PLAN

50 Head Beef/Cattle Corral

• Details of Blocking Gate

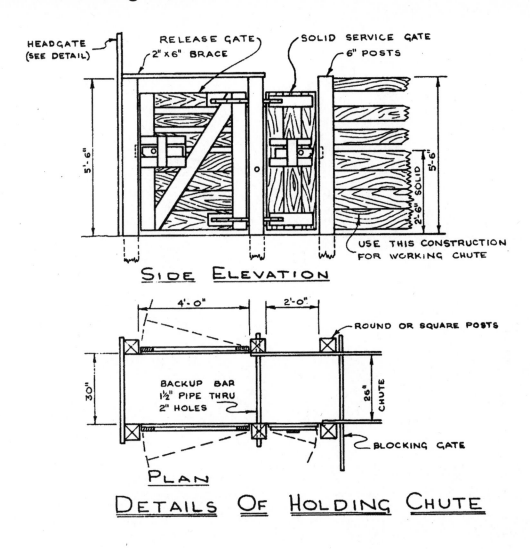

HEADGATE (SEE DETAIL)

RELEASE GATE
2" X 6" BRACE

SOLID SERVICE GATE
6" POSTS

5'-6"

5'-6"

2'-6" SOLID

USE THIS CONSTRUCTION FOR WORKING CHUTE

SIDE ELEVATION

4'-0"

2'-0"

ROUND OR SQUARE POSTS

30"

BACKUP BAR
1½" PIPE THRU
2" HOLES

26" CHUTE

BLOCKING GATE

PLAN

DETAILS OF HOLDING CHUTE

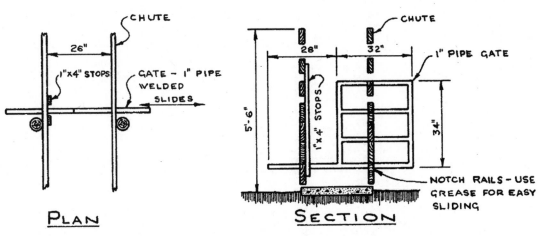

CHUTE

26"

1"X4" STOPS

GATE - 1" PIPE
WELDED
SLIDES

PLAN

CHUTE

28"

32"

1" PIPE GATE

5'-6"

1"X4" STOPS

34"

NOTCH RAILS - USE GREASE FOR EASY SLIDING

SECTION

DETAILS OF BLOCKING GATE

- Headgate Front View • Headgate Anchor Bracket
- Headgate Perspective View

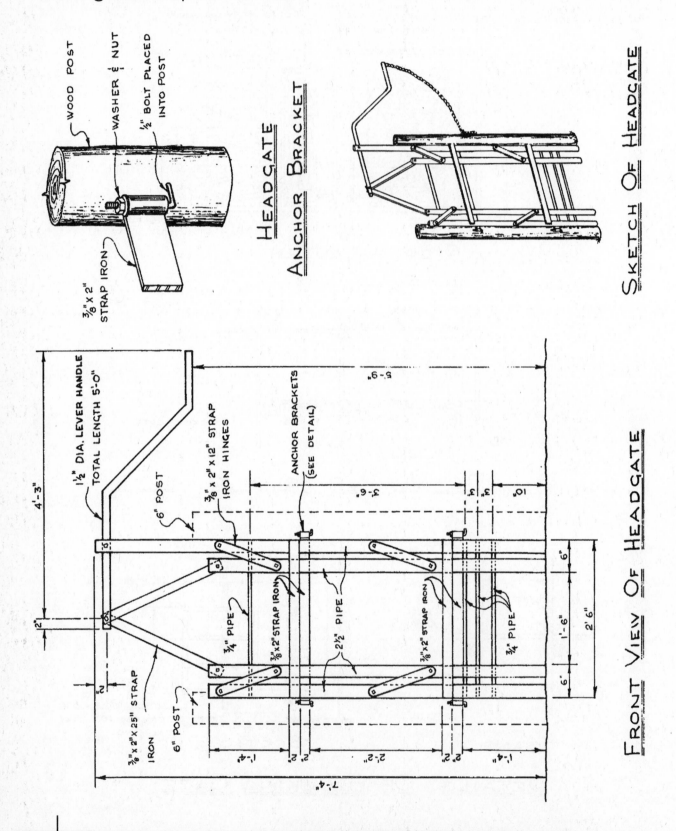

WOOD POST

WASHER & NUT

½" BOLT PLACED INTO POST

⅜" x 2" STRAP IRON

HEADGATE ANCHOR BRACKET

SKETCH OF HEADGATE

1½" DIA. LEVER HANDLE TOTAL LENGTH 5'-0"

4'-3"

6" POST

⅜" x 2" x 12" STRAP IRON HINGES

ANCHOR BRACKETS (SEE DETAIL)

5'-9"

3'-6"

3"

3"

10"

2"

¾" PIPE

⅜" x 2" STRAP IRON

2½" PIPE

⅜" x 2" STRAP IRON

¾" PIPE

6"

1'-6"

2'-6"

6"

FRONT VIEW OF HEADGATE

2"

⅜" x 2" x 25" STRAP IRON

6" POST

1'-4"

2"

2'-2"

2"

1'-4"

7'-4"

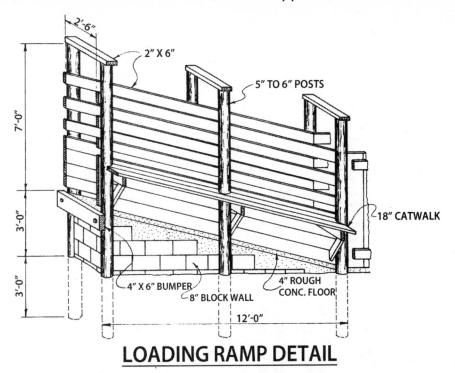

LOADING RAMP DETAIL

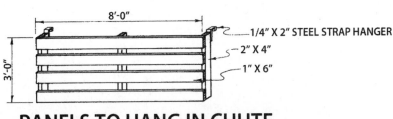

PANELS TO HANG IN CHUTE
FOR WORKING SMALL CALVES

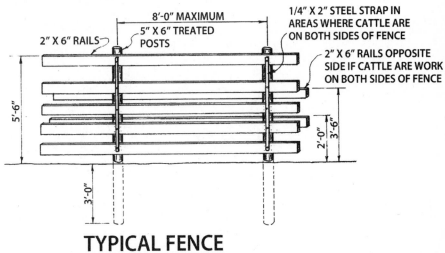

TYPICAL FENCE

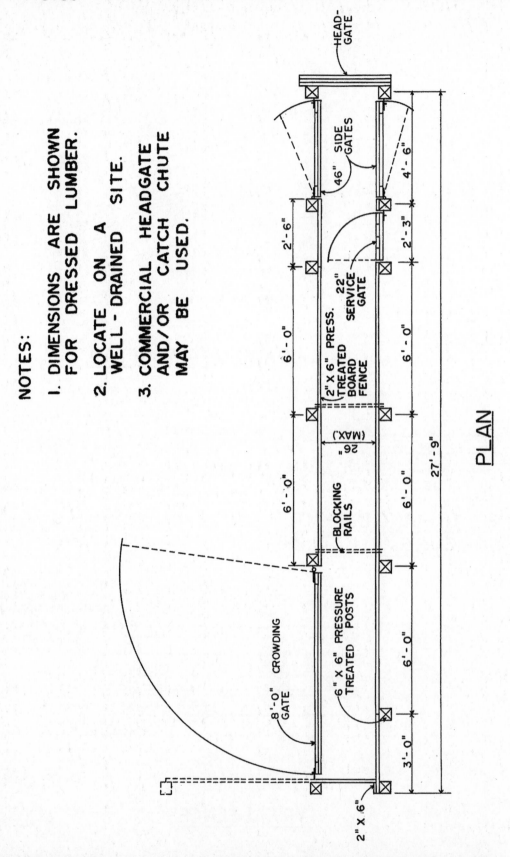

NOTES:

1. DIMENSIONS ARE SHOWN FOR DRESSED LUMBER.

2. LOCATE ON A WELL-DRAINED SITE.

3. COMMERCIAL HEADGATE AND/OR CATCH CHUTE MAY BE USED.

HEAD-GATE

46" SIDE GATES

4'-6"

2'-6"

2'-3"

2" X 6" PRESS. TREATED BOARD FENCE

22" SERVICE GATE

6'-0"

6'-0"

26" (MAX.)

BLOCKING RAILS

6'-0"

6'-0"

27'-9"

8'-0" CROWDING GATE

6" X 6" PRESSURE TREATED POSTS

6'-0"

3'-0"

2" X 6"

PLAN

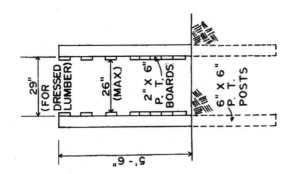

CROSS SECTION

29" (FOR DRESSED LUMBER)

26" (MAX.)

2" X 6" P. T. BOARDS

6" X 6" P. T. POSTS

5' - 6"

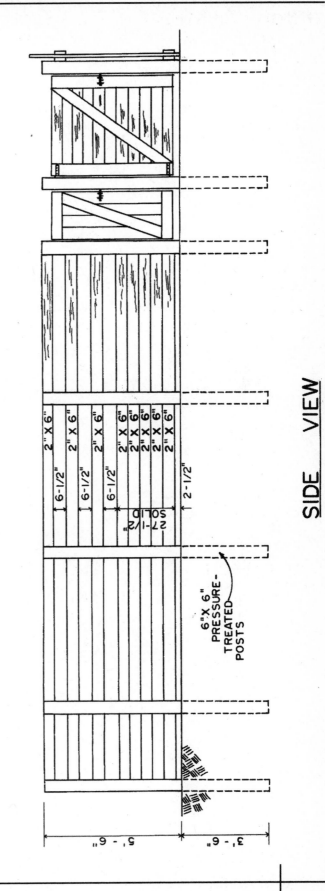

SIDE VIEW

2" X 6"
2" X 6"
2" X 6"
2" X 6"
2" X 6"
2" X 6"
2" X 6"
2" X 6"

6-1/2"
6-1/2"
6-1/2"

27-1/2" SOLID

2-1/2"

6" X 6" PRESSURE-TREATED POSTS

5' - 6"

3' - 6"

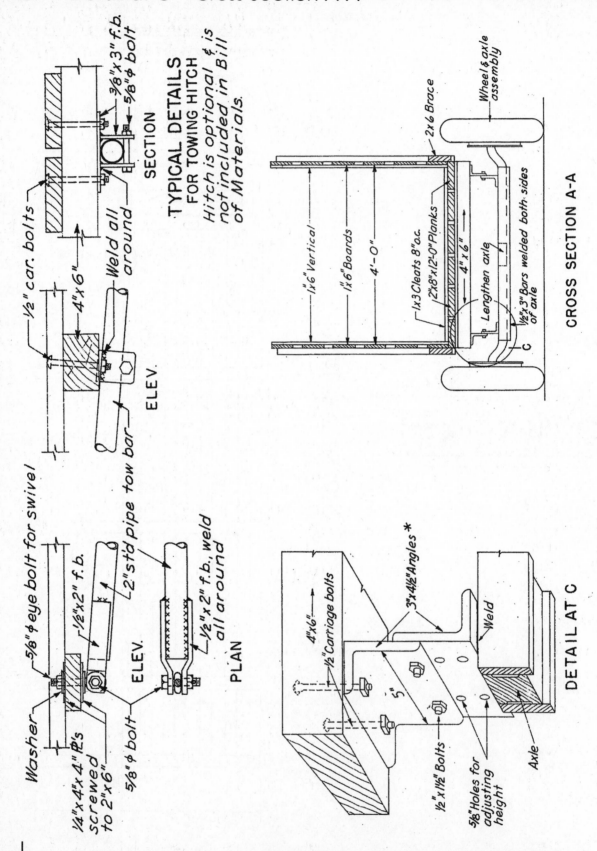

MOVEABLE CHUTE FOR LOADING CATTLE

• Longituinal Section • Detail at D

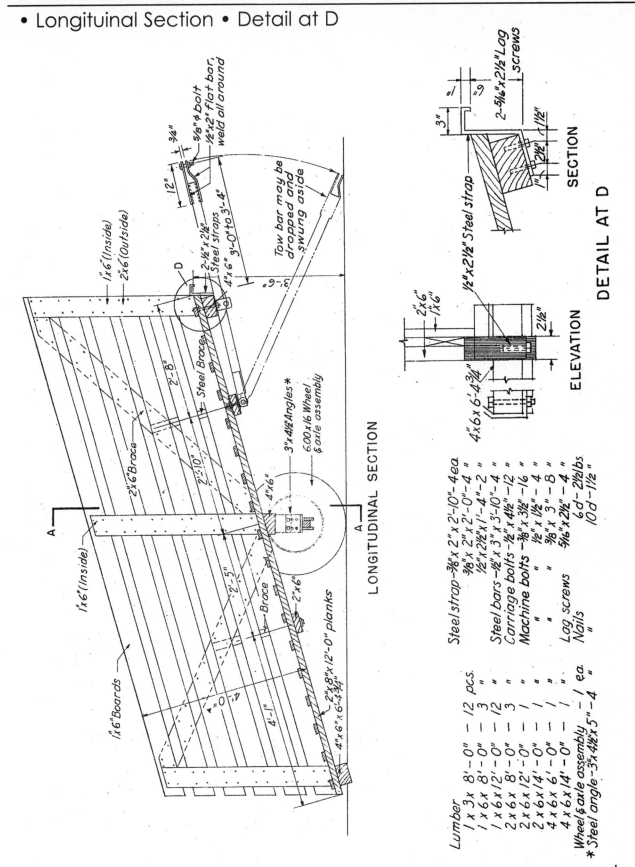

LONGITUDINAL SECTION

DETAIL AT D

SECTION

ELEVATION

Lumber
1 x 3 x 8'-0" — 12 pcs.
1 x 6 x 8'-0" — 3 "
1 x 6 x 12'-0" — 12 "
2 x 6 x 8'-0" — 3 "
2 x 6 x 12'-0" — 1 "
2 x 6 x 14'-0" — 1 "
4 x 6 x 6'-0" — 1 "
4 x 6 x 14'-0" — 1 "
Wheel & axle assembly — 1 ea.
* Steel angle - 3"x 4½"x 5" — 4 "

Steel strap - ⅜" x 2" x 2'-10" - 4 ea.
⅜" x 2" x 2'-0" - 4 "
½" x 2½" x 1'-4" - 2 "
Steel bars - ½" x 3" x 3'-10" - 4 "
Carriage bolts - ½" x 4½" - 12 "
Machine bolts - ⅜" x 3½" - 16 "
½" x 1½" - 4 "
⅜" x 3 - 8 "
Lag screws 5/16" x 2½" - 4 "
Nails 6d - 2½ lbs.
10d - 1½ "

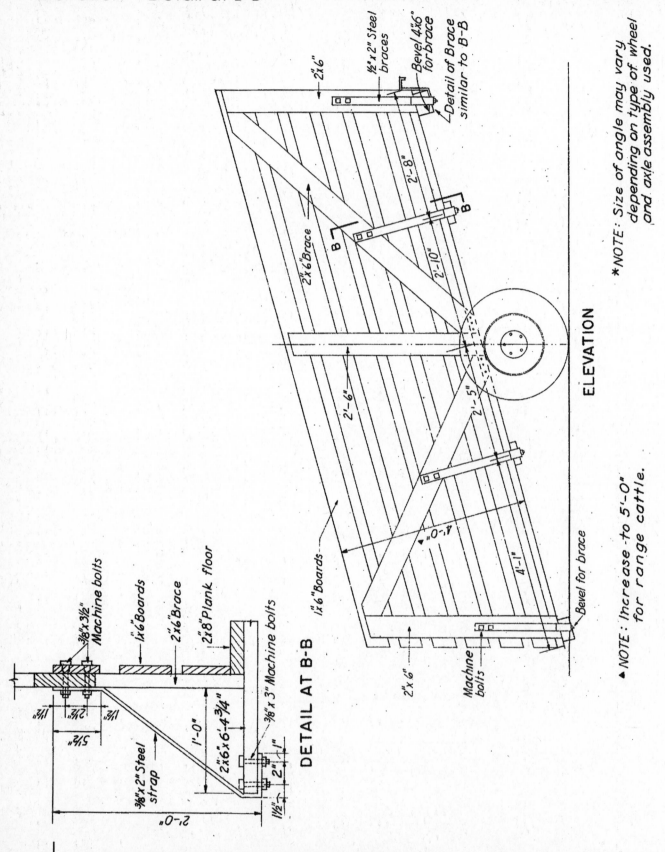

2"x6"

½"x 2" Steel braces

Bevel 4½"x6" for brace

Detail of Brace similar to B-B

2'-8"

B

2'-10"

B

2"x6"Brace

2'-6"

2'-5"

4'-0"

1"x6"Boards

4'-1"

Bevel for brace

2"x6"

Machine bolts

ELEVATION

*NOTE : Size of angle may vary depending on type of wheel and axle assembly used.

▲ NOTE : Increase to 5'-0" for range cattle.

⅜"x 3½" Machine bolts

1"x6"Boards

2"x6"Brace

2x8"Plank floor

⅜"x 3" Machine bolts

2"x6x6'-4¾"

⅜"x 2" Steel strap

1'-0"

1½" 2½" 1½"

5½"

2'-0"

1½" 2" 1"

DETAIL AT B-B

160

THE COMPLETE GUIDE TO BUILDING CLASSIC BARNS, FENCES, STORAGE SHEDS, ANIMAL PENS, OUTBUILDINGS, GREENHOUSES, FARM EQUIPMENT, & TOOLS

• Plan

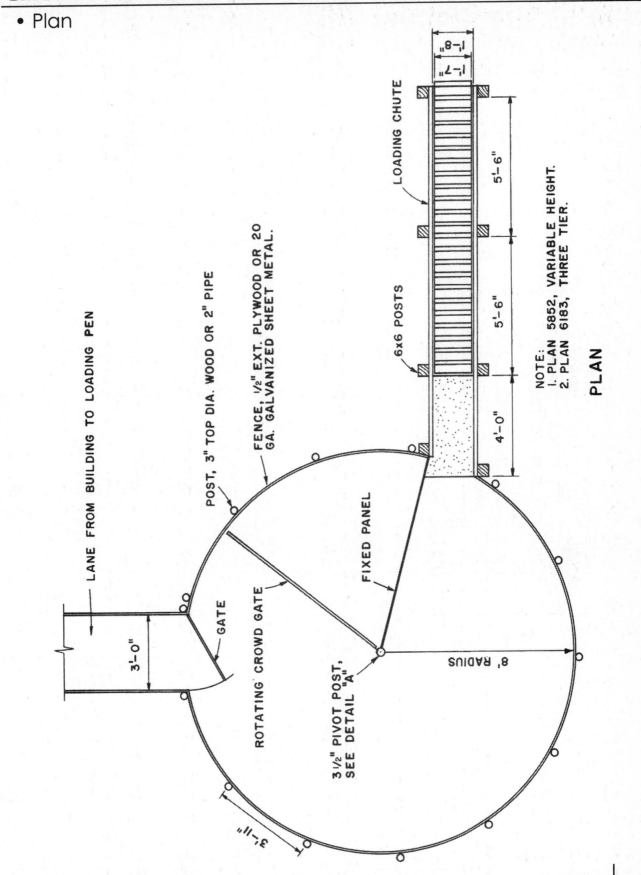

LANE FROM BUILDING TO LOADING PEN

POST, 3" TOP DIA. WOOD OR 2" PIPE

FENCE, ½" EXT. PLYWOOD OR 20 GA. GALVANIZED SHEET METAL.

LOADING CHUTE

6x6 POSTS

1'-8"

1'-7"

5'-6"

5'-6"

4'-0"

NOTE:
1. PLAN 5852, VARIABLE HEIGHT.
2. PLAN 6183, THREE TIER.

PLAN

FIXED PANEL

ROTATING CROWD GATE

GATE

3½" PIVOT POST, SEE DETAIL "A"

8' RADIUS

3'-0"

3'-11"

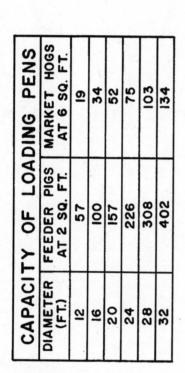

CAPACITY OF LOADING PENS

DIAMETER (FT.)	FEEDER PIGS AT 2 SQ. FT.	MARKET HOGS AT 6 SQ. FT.
12	57	19
16	100	34
20	157	52
24	226	75
28	308	103
32	402	134

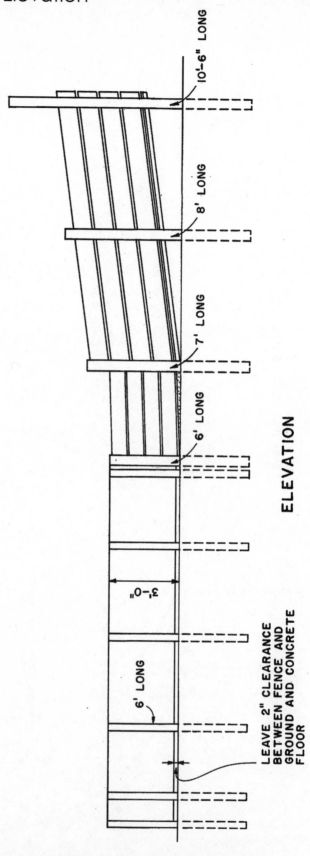

ELEVATION

• Detail A

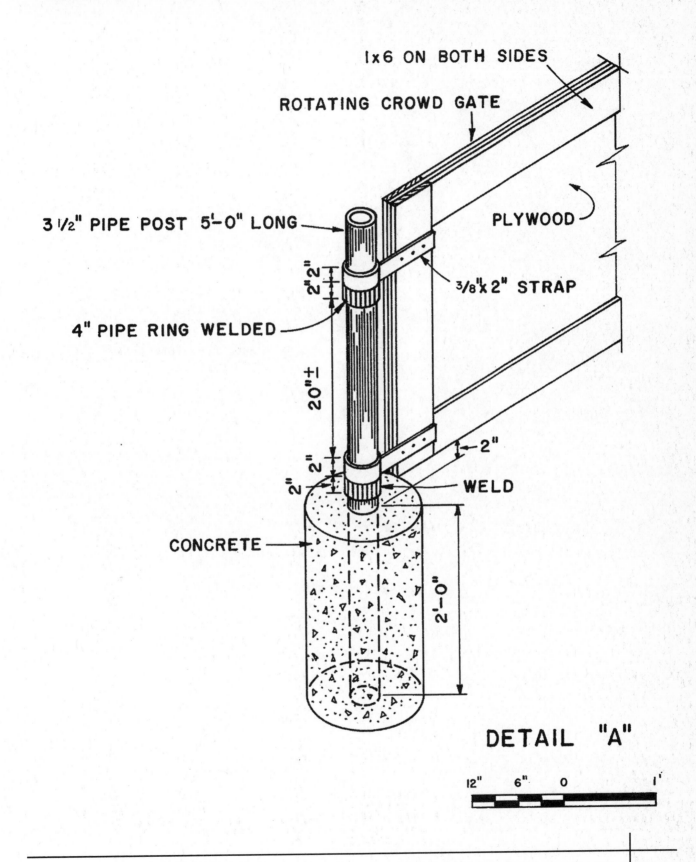

1x6 ON BOTH SIDES

ROTATING CROWD GATE

PLYWOOD

3 1/2" PIPE POST 5'-0" LONG

2"2"

3/8" x 2" STRAP

4" PIPE RING WELDED

20"±

2"
2"
2"

2"

WELD

CONCRETE

2'-0"

DETAIL "A"

12"   6"   0   1'

THE COMPLETE GUIDE TO BUILDING CLASSIC BARNS, FENCES, STORAGE SHEDS, ANIMAL PENS, OUTBUILDINGS, GREENHOUSES, FARM EQUIPMENT, & TOOLS

chapter 10

CATTLE & AUCTION BARN

• Perspective View

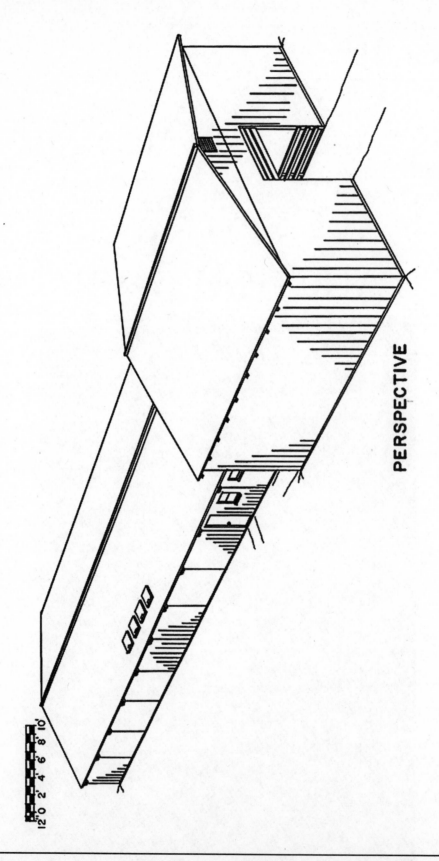

PERSPECTIVE

• Floor Plan

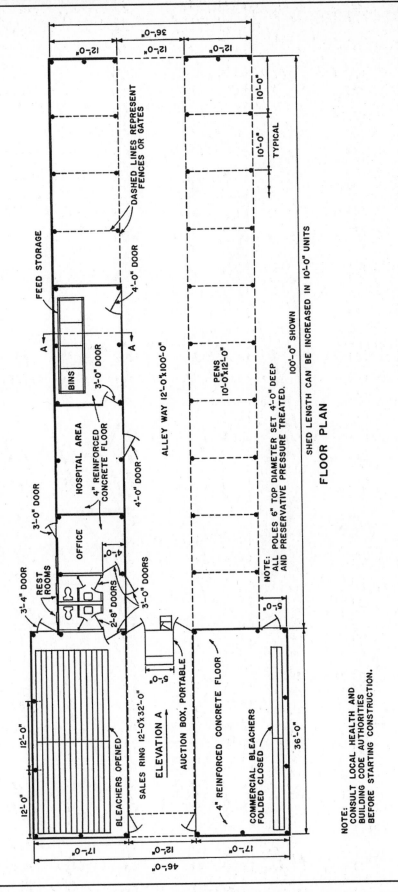

FLOOR PLAN

• Auctioneer's Box

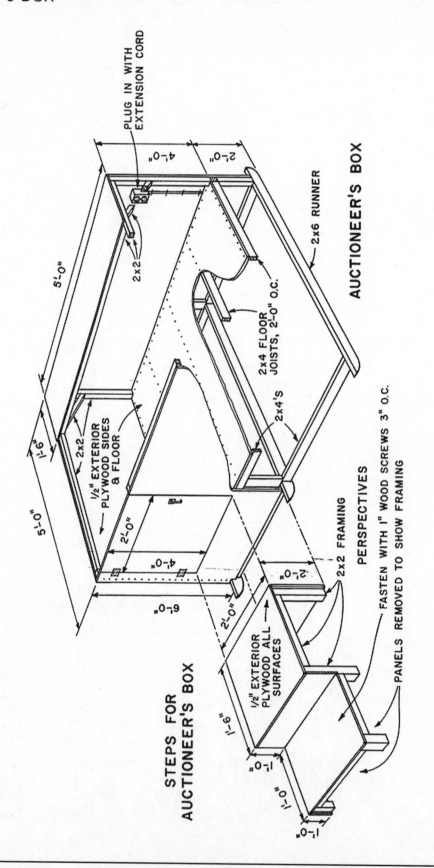

PLUG IN WITH EXTENSION CORD

4'-0"

2'-0"

AUCTIONEER'S BOX

2x6 RUNNER

2x2

5'-0"

2x4 FLOOR JOISTS, 2'-0" O.C.

2x4'S

1'-6"

2x2

1/2" EXTERIOR PLYWOOD SIDES & FLOOR

PERSPECTIVES

2x2 FRAMING

FASTEN WITH 1" WOOD SCREWS 3" O.C.

PANELS REMOVED TO SHOW FRAMING

5'-0"

2'-0"

4'-0"

6'-0"

2'-0"

2'-0"

STEPS FOR AUCTIONEER'S BOX

1'-6"

1/2" EXTERIOR PLYWOOD ALL SURFACES

1'-0"

1'-0"

1'-0"

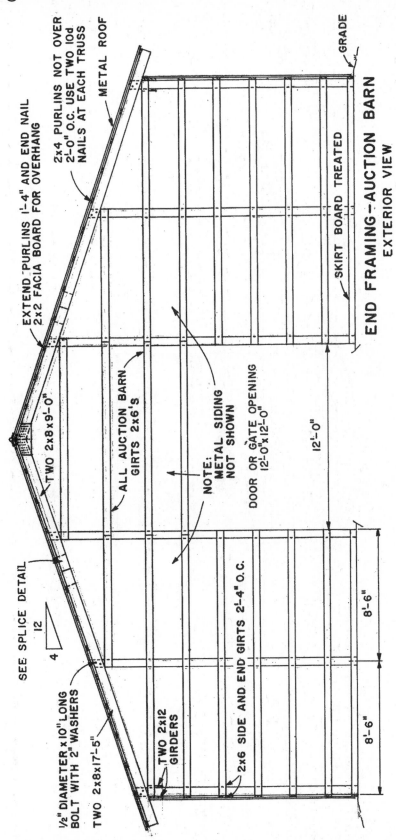

END FRAMING - AUCTION BARN
EXTERIOR VIEW

GRADE

METAL ROOF

2x4 PURLINS NOT OVER 2'-0" O.C. USE TWO 10d NAILS AT EACH TRUSS

EXTEND PURLINS 1'-4" AND END NAIL 2x2 FACIA BOARD FOR OVERHANG

TWO 2x8x9'-0"

SEE SPLICE DETAIL

12
4

1/2" DIAMETER x10" LONG BOLT WITH 2" WASHERS

TWO 2x8x17'-5"

ALL AUCTION BARN GIRTS 2x6'S

NOTE: METAL SIDING NOT SHOWN

DOOR OR GATE OPENING 12'-0"x12'-0"

SKIRT BOARD TREATED

12'-0"

8'-6"

8'-6"

TWO 2x12 GIRDERS

2x6 SIDE AND END GIRTS 2'-4" O.C.

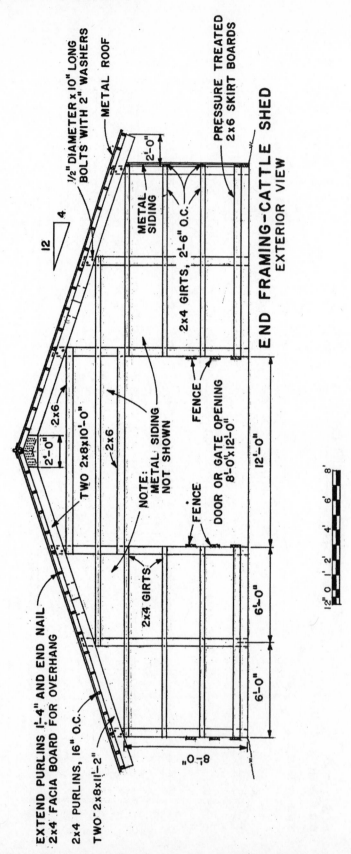

END FRAMING–CATTLE SHED
EXTERIOR VIEW

CATTLE & AUCTION BARN

• Framing Between Cattle Shed & Auction Barn

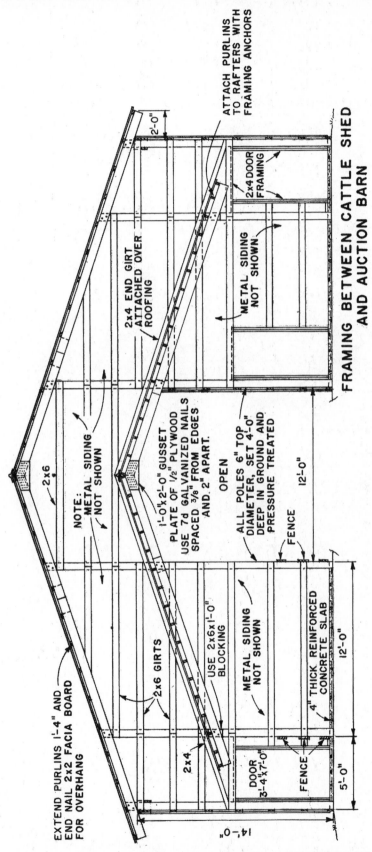

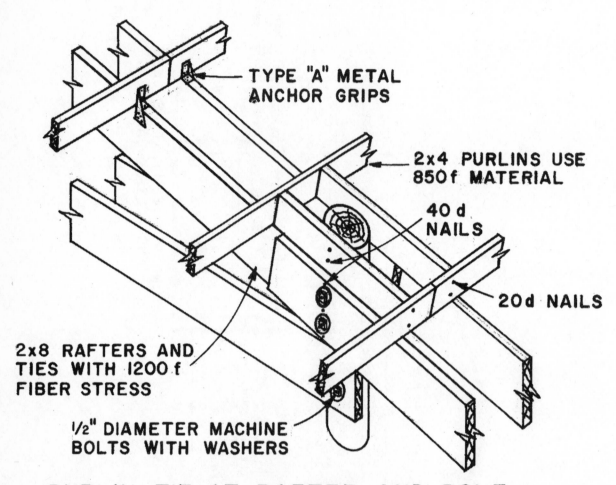

TYPE "A" METAL ANCHOR GRIPS

2x4 PURLINS USE 850 f MATERIAL

40 d NAILS

20 d NAILS

2x8 RAFTERS AND TIES WITH 1200 f FIBER STRESS

1/2" DIAMETER MACHINE BOLTS WITH WASHERS

PURLIN TIE AT RAFTER AND POLE
PERSPECTIVE

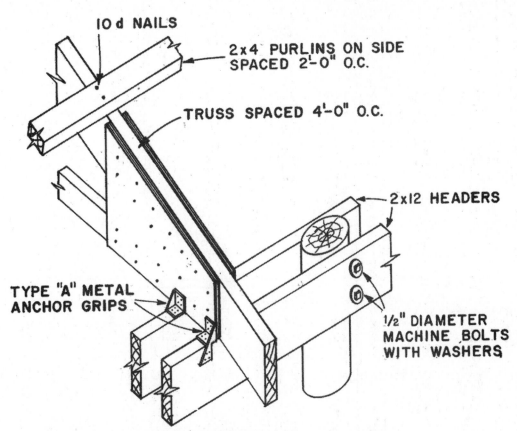

10d NAILS

2x4 PURLINS ON SIDE
SPACED 2'-0" O.C.

TRUSS SPACED 4'-0" O.C.

2x12 HEADERS

TYPE "A" METAL
ANCHOR GRIPS

½" DIAMETER
MACHINE BOLTS
WITH WASHERS

TRUSS ATTACHMENT TO HEADER
ISOMETRIC

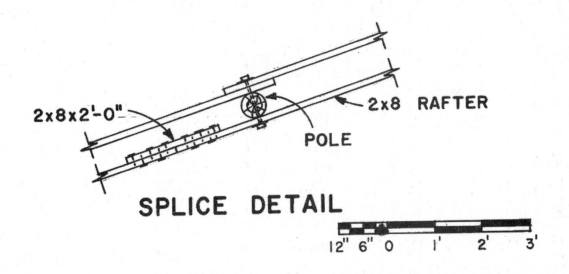

2x8x2'-0"

2x8 RAFTER

POLE

SPLICE DETAIL

12" 6" 0 1' 2' 3'

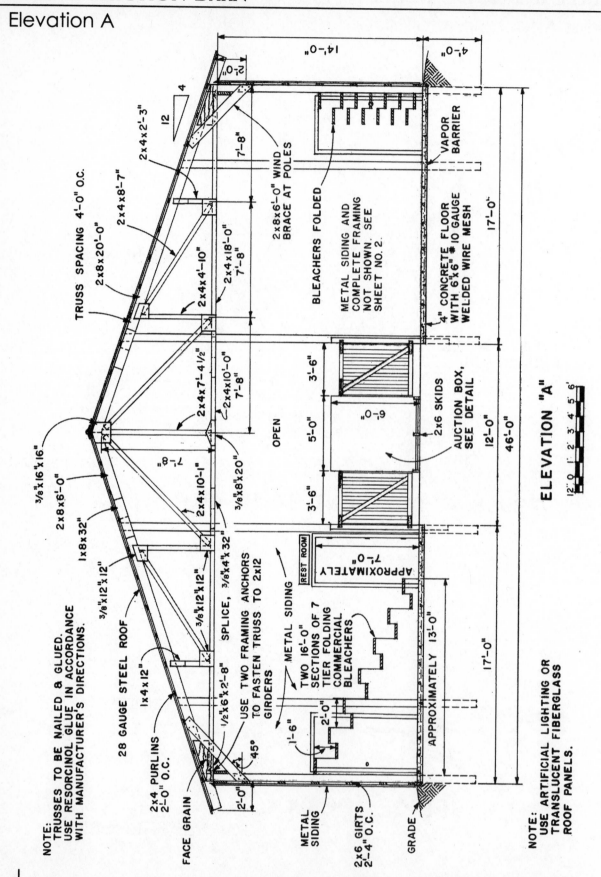

• Feed Bins

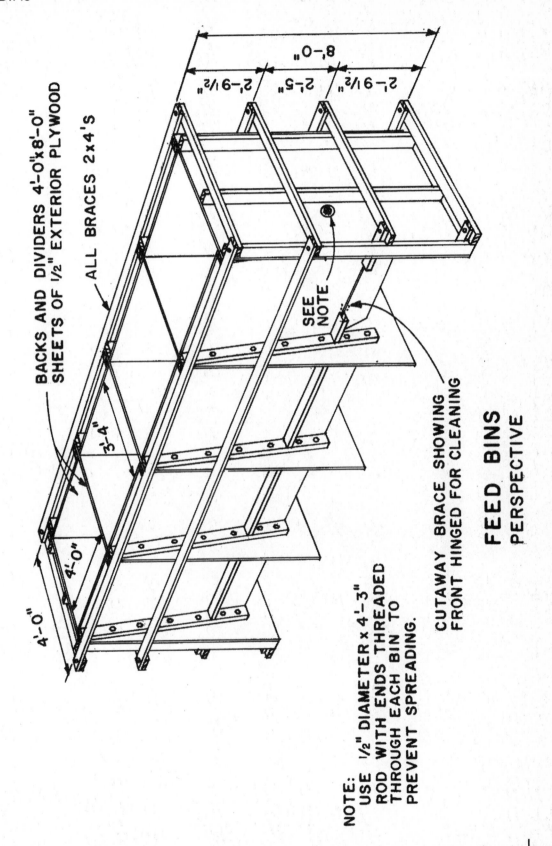

BACKS AND DIVIDERS 4'-0"x8'-0"
SHEETS OF 1/2" EXTERIOR PLYWOOD

ALL BRACES 2x4'S

8'-0"

2'-9 1/2"   2'-5"   2'-9 1/2"

SEE NOTE

3'-4"

4'-0"

4'-0"

FEED BINS
PERSPECTIVE

CUTAWAY BRACE SHOWING
FRONT HINGED FOR CLEANING

NOTE:
USE 1/2" DIAMETER x 4'-3"
ROD WITH ENDS THREADED
THROUGH EACH BIN TO
PREVENT SPREADING.

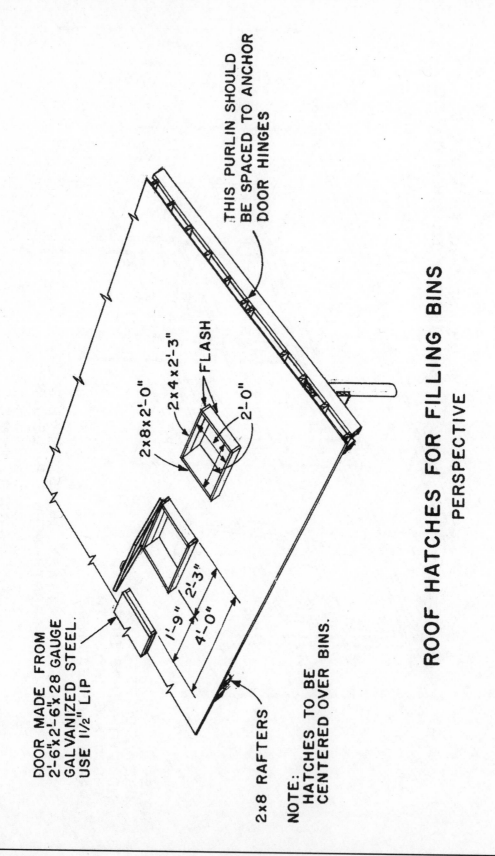

THIS PURLIN SHOULD BE SPACED TO ANCHOR DOOR HINGES

2x8x2'-0"

2x4x2'-3"

FLASH

2'-0"

1'-9" 2'-3"

4'-0"

DOOR MADE FROM 2'-6"x2'-6"x 28 GAUGE GALVANIZED STEEL. USE 1½" LIP

2x8 RAFTERS

NOTE: HATCHES TO BE CENTERED OVER BINS.

ROOF HATCHES FOR FILLING BINS
PERSPECTIVE

• Section A-A of Feed Storage Bins

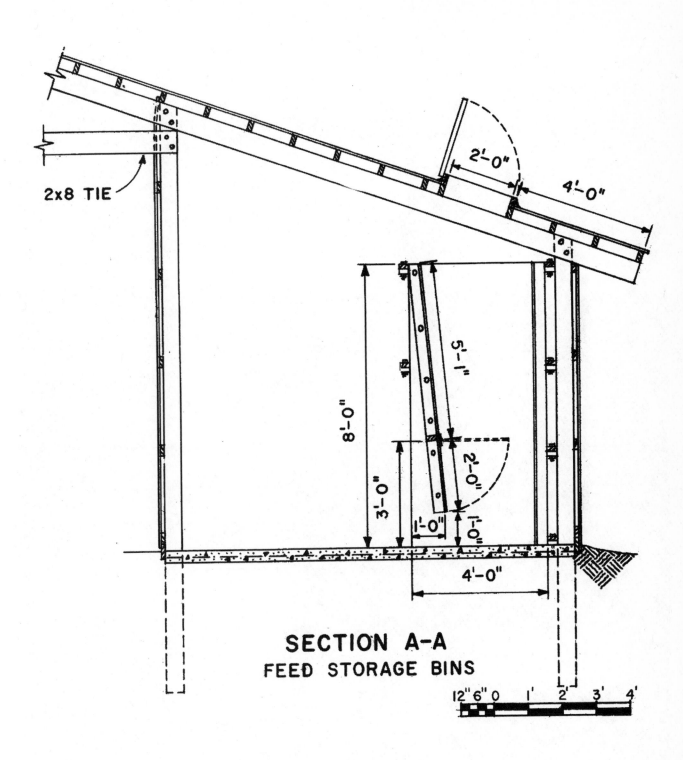

2x8 TIE

2'-0"

4'-0"

5'-1"

8'-0"

3'-0"

2'-0"

1'-0"

1'-0"

4'-0"

SECTION A-A
FEED STORAGE BINS

12" 6" 0   1'   2'   3'   4'

The Complete Guide to Building Classic Barns, Fences, Storage Sheds, Animal Pens, Outbuildings, Greenhouses, Farm Equipment, & Tools

chapter 11

GREENHOUSES

5x8 Lean-To Greenhouse

• Perspective • Bill of Material

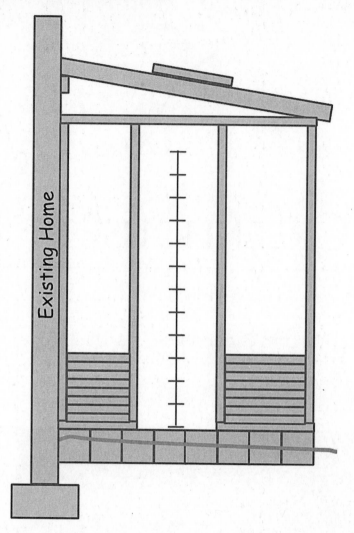

Bill of Material

Item	Qty	Description
1	8	8x8 timber x 8'-0"
2	6	2x4 x 5'-9"
3	1	2x4 x 8'-0"
4	2	2x4 x 7'-5"
5	13	2x4 x 6'-0"
6	1 box	3" long 10d nails
7	2	2x4 x 1'-9"
8	2	2x4 x 2'-0"

Item	Qty	Description
9	2	2x4 x 5'-4"
10	2	2x4 x 1'-5"
11	2	2x4 x 1'-8"
12	1	roof vent
13	2	zippers
14	2	wall vents
15	1 roll	UV resistant polyethelyne

• Top View & End View Timber Foundation

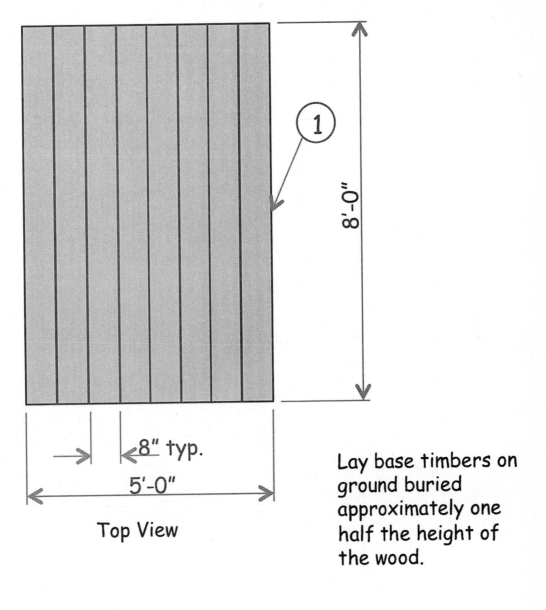

8'-0"

1

8" typ.

5'-0"

Top View

Lay base timbers on ground buried approximately one half the height of the wood.

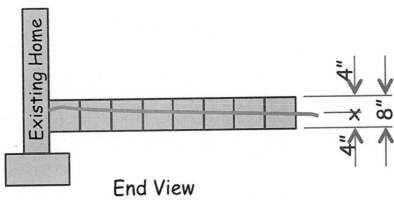

Existing Home

4"

x

8"

4"

End View

• Top & End View of Frame

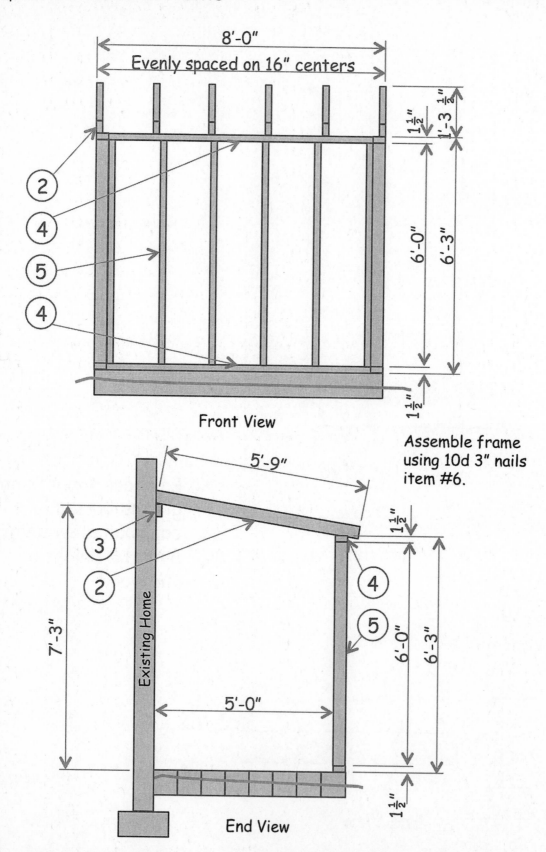

8'-0"

Evenly spaced on 16" centers

② ④ ⑤ ④

1 1/2"   1 1/2"

1'-3"

6'-0"   6'-3"

1 1/2"

Front View

Assemble frame using 10d 3" nails item #6.

5'-9"

③ ② ④ ⑤

Existing Home

7'-3"

5'-0"

1 1/2"

6'-0"   6'-3"

1 1/2"

End View

5x8 Lean-To Greenhouse

• End Frame

Assemble end frame
using 10d 3" nails
item #6.

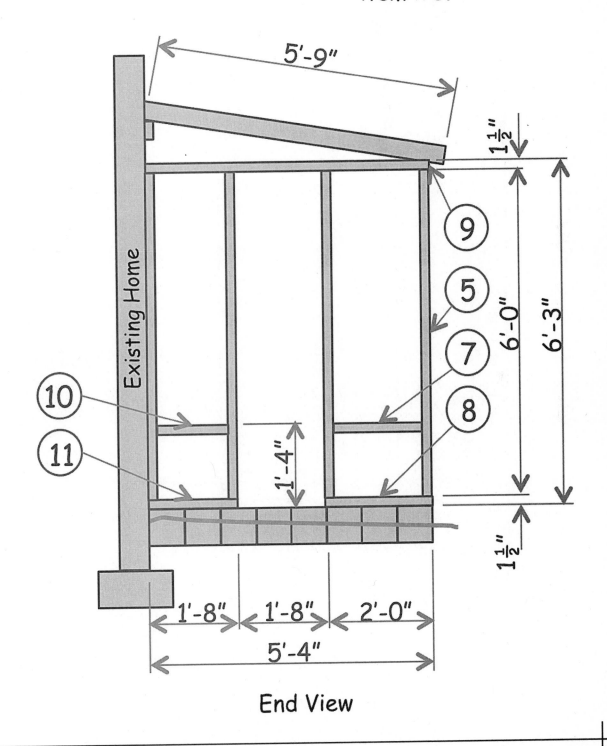

End View

• Final Assembly

Install vents and cover structure
with UV resistant polyethylene or
fiberglass panels.

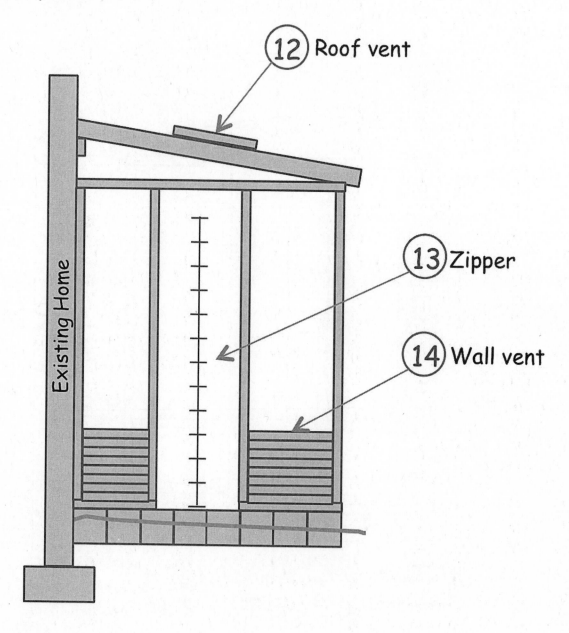

(12) Roof vent

(13) Zipper

(14) Wall vent

Existing Home

End View

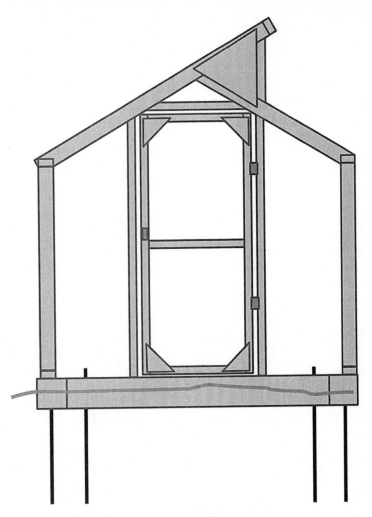

Bill of Material

Item	Qty	Description	Item	Qty	Description
1	1	8x8 timber x 6'-0"	11	1 roll	UV resistent poylethylene
2	2	8x8 timber x 10'-0"	12	6	1/2" plywood x 1'-3" x 1'-9"
3	10	1/2" dia. rebar x 3'-0" LG	13	3	2x2 x 2'-2"
4	5	2x4 x 10'-0"	14	2	2x2 x 5'-9"
5	6	2x4 x 6'-0"	15	4	2x4 x 5'-10"
6	6	2x4 x 4'-0"	16	4	1/2" plywood x 6" x 6"
7	12	2x4 x 4'-6"	17	2	hinges
8	6	2x4 x 1'-6"	18	1	handle
9	1 box	10d nails x 3" LG	19	1 box	#8 wood screws x 2 1/2" LG
10	20	1x4 x 1'-10 1/4"	20	1 box	#8 wood screws x 1 1/2" LG

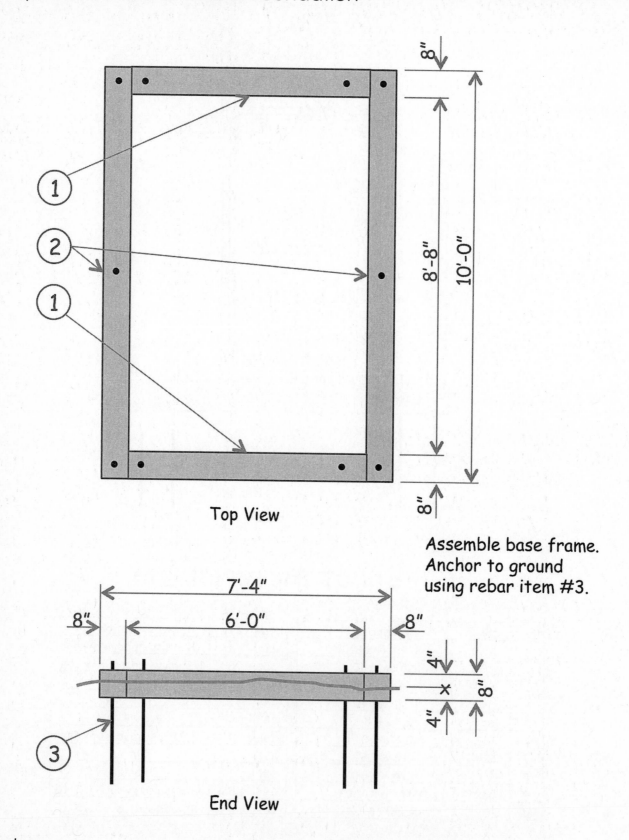

Top View

Assemble base frame.
Anchor to ground
using rebar item #3.

End View

7x10 A-Frame Greenhouse

• Side & End View of Frame

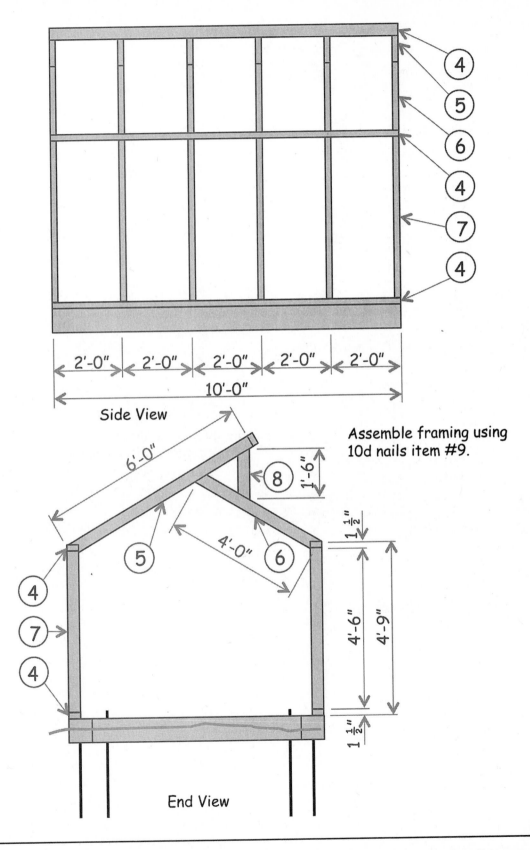

Side View

Assemble framing using
10d nails item #9.

End View

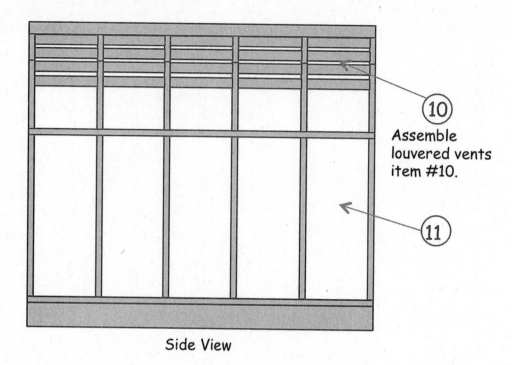

Assemble
louvered vents
item #10.

Side View

1'-3"

1'-9"

2'-2"

5'-9"

Assemble door , install
gussets, and cover
structure with UV
resistant polyethylene.

End View

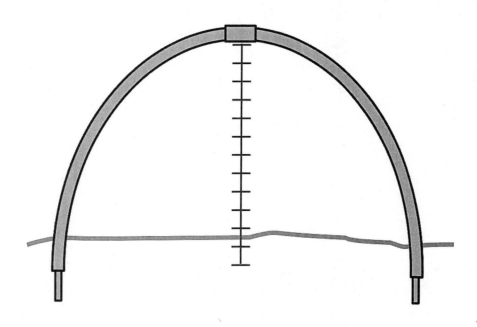

Bill of Material

Item	Qty	Description	Item	Qty	Description
1	16	1/2" PVC pipe x 3'-0"	5	1	Zipper
2	6	3/4" cross tee	6	1 roll	UV resistant polyethylene
3	16	3/4" PVC pipe x 10'-0"	7	48	3/4" PVC x 6"
4	2	3/4" tee	8	1 can	PVC cement

• Top & Front Views of Support Stakes

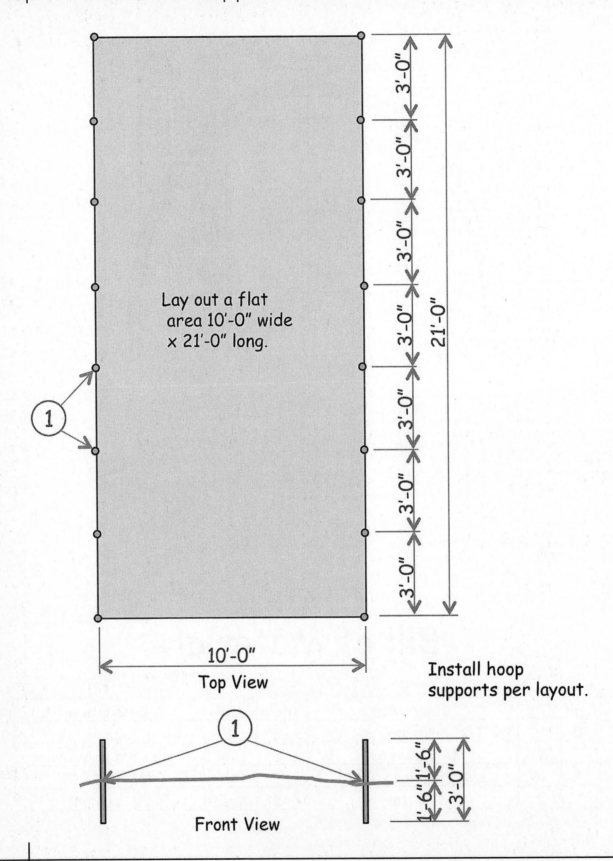

Lay out a flat area 10'-0" wide x 21'-0" long.

①

3'-0"
3'-0"
3'-0"
3'-0"
3'-0"
3'-0"
3'-0"

21'-0"

10'-0"

Top View

Install hoop supports per layout.

①

Front View

1'-6" 1'-6"
3'-0"

• Top View of Hoops

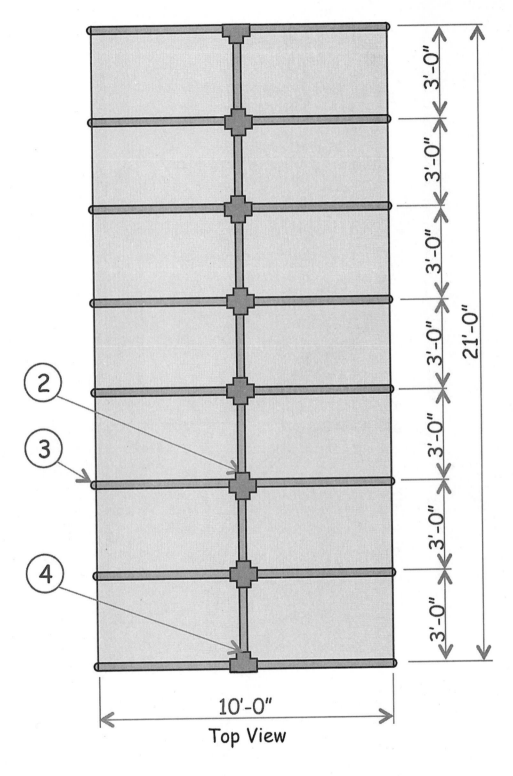

21'-0"

3'-0"
3'-0"
3'-0"
3'-0"
3'-0"
3'-0"
3'-0"

Install hoops
into cross tees.

2

3

4

10'-0"

Top View

• Final Assembly

Split item #7 and snap in place over hoop to contain polyethylene cover.

Enlarged view of item #7.

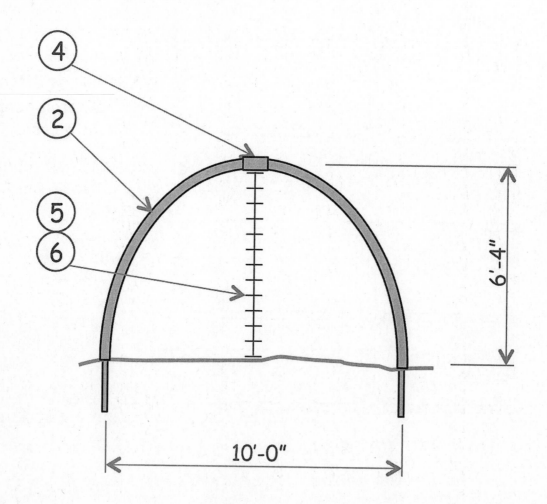

④

②

⑤

⑥

6'-4"

10'-0"

Front View